For Mum,
the inspiration behind my first 'Fun Science' video.
(Thanks for the great genes.)

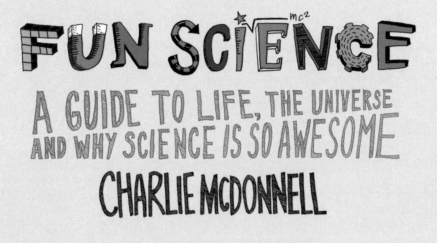

FUN SCIENCE

A GUIDE TO LIFE, THE UNIVERSE AND WHY SCIENCE IS SO AWESOME

CHARLIE MCDONNELL

quadrille

*Charlie McDonnell does not take responsibility for any existential crises that this book may cause. The universe is a big, big place man – it's not his fault if it makes you feel insignificant.

Publishing Director Sarah Lavelle
Junior Commissioning Editor Romilly Morgan
Creative Director Helen Lewis
Design & Photography Dave Brown, apeinc.co.uk
Illustration Fran Meneses
Design Assistant Gemma Hayden
Cover Illustration Ivana Zorn
Production Tom Moore, Vincent Smith

First published in 2016 by
Quadrille Publishing
Pentagon House
52–54 Southwark Street
London SE1 1UN
www.quadrille.co.uk
www.quadrille.com

Reprinted in 2016
10 9 8 7 6 5 4 3 2

Quadrille is an imprint of Hardie Grant
www.hardiegrant.com.au

Cataloguing in Publication Data: a catalogue record
for this book is available from the British Library.

ISBN: 978 184949 802 9

Printed in China

CONTENTS

INTRODUCTION

||||||||||||||||||||||||||||||||||||||||||||||||||||||||||||||||||||||||||||||||||||||||

Hello! I'm Charlie, and before we get started...
I need to be honest with you:

I AM NOT A

||||||||||||||||||||||||||||||||||||||||||||||||||||||||||||||||||||||||||||||||||||||||

SCIENTIST.

||||||||||||||||||||||||||||||||||||||||||||||||||||||||||||||||||||||||||||||||||||||||

Right now you're in a bookstore, or you've dipped into the preview pages
of this book online, or maybe you've already bought *Fun Science,* and you're
only now remembering that this Charlie McDonnell guy is ultimately just
another YouTuber with a book deal, and how much can he really teach you
about science given that his major scientific qualifications are the As he got
in his school exams.

Still bragging about those, apparently.

Basically, you're judging me... and I get it. I really do. Because ultimately
you're only judging me because I'm judging me, and you're just a person who
exists inside my head (...which means you're judgemental of me by default).

HOWEVER, this is supposed to be a science book...

Before I attempt to make my case as to why I think non-scientists can be just as good (if not possibly better... maybe... very maybe) at sharing the best and most interesting scientific knowledge with the world, let me get you in the mood for some learning by answering one, seemingly very simple question:

WHAT IS SCIENCE?

Before we go any further, just take a moment to try and answer that question. If you're already well-versed in the wonders of science then it's probably a pretty easy one for you, but if your interest in the topic is still relatively new you might find it slightly harder than you'd expect.

And please, don't Google it. **RESIST**. Just for a moment...

So, have you got an answer in your head? Well, if you're anything like I was in my *'I guess science is fine or whatever'* years, then you might respond like this: **science is simply another group of things that you had to learn in order to not be in school anymore.**

When I thought of science, I used to think of it as a subject. It was a means to a grade, another lesson in my timetable. It was an opportunity to see something blow up... and then usually forget what it was I was meant to have learnt from the explosion. It was a time to nervously file away as much information as I could in case, god forbid, my biology teacher decided to pick on me for an answer, so that I didn't embarrass myself (at least in class). Well, either it was just a subject, or it was that thing that I heard about in the news – something that other, smarter people made significant breakthroughs in within their area of the field.

And sadly, to many people, this is all science ever is to them. At least, that's my hunch. To be honest, I wasn't entirely sure of how the majority of people perceived science. So, instead of just trying to pass off my hunch as a fact, I did a bit of my own research – by asking you guys!

Over 500 people, if you're interested.

Once upon a time I had planned to make a video on this very topic: 'What is science?', so I decided to ask my online audience that very question. Of all the answers I received, one response that came up time and time again was something along the lines of what @anonymous1 said:

Science is... *"a boring subject which I am forced to suffer through at school."*

In a sense, @anonymous1, you're not entirely wrong. The beauty of language is that words mean what we choose them to mean. So, if science is just a subject to you, then you're ultimately right. Like you, there were also many times when the science I learnt at school seemed to be incredibly boring and uninspiring (for whatever reason). I only really started to develop a passion for science after I'd left school and started teaching myself about it.

But the definition I was searching for, the dictionary definition of science, is much more inspiring. Here's another answer from @anonymous2:

Science is... *"a way of looking at and explaining our world. Trying to explain HOW things happen. Science is continually evolving with new theories created all the time, through never-ending 'trial and error' and methodical experimenting."*

Frankly, I couldn't have put it better myself! (Although I am now, of course, going to try to do just that.)

What @anonymous2 was describing is science as a method for figuring out how the world works, and as the body of all knowledge obtained from this very particular method.

It's a shame then that when you think of 'the scientific method', you might end up picturing something like this:

OBSERVATION
HYPOTHESIS
EXPERIMENT
ANALYSIS
CONCLUSION

Five words which, in that particular order, inspire an acute sense of boredom and exhaustion (in me at least!). If counting sheep doesn't do it for you, reading those words over and over again should undoubtedly do the trick.

But that feeling of boredom arises because we were taught to use these words in a certain way. To many, they're not a guide to learning how the world works – they're a checklist of work we have to do in order to pass a class! Which means that, really, we need to change our attitude towards them.

We need some new words. How about:

...I did my best.

ALRIGHT, maybe we should use the real words, but my new ones can just help to inspire the feeling of how awesome the scientific method is. **(But like, a good feeling, you know? Not an "oh my God, Charlie's puns are so bad" kind of feeling.)**

And you really, really should be excited about the scientific method, because it is absolutely the **BEST** tool that we have for figuring out the *truth*. Heck, we've got the proof of how well it works all around us: from the rectangular computers in our pockets to the electricity in our homes, all the way to the knowledge about what those sparkly little lights in the sky actually are. So much of what we do and know to be true about our everyday, modern life wouldn't be possible if it weren't for the vast body of scientific knowledge that humankind has been collecting for thousands of years.

So how, pray tell, does this method actually work? Well, as an example, I actually already started using the scientific method earlier in this introduction... when I was wondering whether or not most people know what science actually is! Here's how it breaks down:

OBSERVATION: This, simply, is when you ask a question that you want to be answered. For me, I noticed that I didn't really understand what science was when I was younger – it was something I had to find out for myself – and so I wondered whether or not other people had a similar experience.

HYPOTHESIS: Before I tried to answer my question, I made a guess: I predicted that most people probably don't think about science as a method for understanding the world, and instead viewed it as a school subject.

EXPERIMENT: Next, I tried to figure out the best way to answer my question. For me, it was a simple case of doing a survey where I asked people what they thought science was, and then put that survey out into the world.

ANALYSIS: Once you're finished with your experiment, it's time to take a look at your results and figure out what they mean. After going through my 500 answers, I found out that 34% of people understood science as being a method for figuring out how the world works.

CONCLUSION: This is where you compare your original guess to what you now know to be true. In my case, I thought that most people probably didn't think of science as a method for understanding the world – and I was right! As many as 66% of the people I asked didn't yet know exactly what science is.

Back in school, if you were asked to put together a little document similar to the one on the previous page (although hopefully yours would be more in-depth) you'd be given a grade, and that'd be that. However, the real beauty of the scientific method comes AFTER you're finished with your experiment...

I mean, let's just think about my survey for a minute (sorry!). Five hundred people isn't a bad size... but would we get a more accurate result if we asked even more people? Or at least, if we asked a more diverse group of people? I, of course, only have access to the folks who follow me on Twitter, and they probably have a slightly more keen interest in science as there's a chance they've seen my *Fun Science* videos in the past. That, and I know the majority of people who watch me are 13–30 years old, so my results are probably a little skewed towards what people in my age range (-ish) think.

Basically, if you take the time to think properly about an experiment once it's finished, you can probably find lots of things that you could have done better. Ultimately, this is a VERY GOOD THING.

Let's just imagine for one moment that we're doing a proper scientific experiment, not just some silly online survey that was created to make a point. What are the steps that happen AFTER you've reached your conclusion?

PEER REVIEW: At this stage I'd share my experiment with a group of really smart scientists, and in the same way I had analyzed the flaws of my own experiment, they'd decide if I had done a good enough job with my survey and whether my test results were worth sharing with scientists all across the world.

PUBLICATION: Assuming I made it past the peer review stage, my entire experiment (all the way from observation to conclusion) would be published in a paper for everyone in the scientific community to see and scrutinize. No longer is it just me judging the quality of my work – EVERYONE is.

REPLICATION: This is probably the best bit. A whole host of people try to either replicate my experiment, or think of better ways to do it, and they go out and collect a load of new data and make their own conclusions... And then the whole cycle starts all over again, and we get closer and closer to the truth!

This really is where the scientific method comes into its own. Science assumes that people are biased and flawed, and that going off the back of what just one person in a book theorizes will not be the best way to find the truth. It also assumes that it doesn't matter how smart the person who came up with their answer is, nor how beautiful and convenient that answer might seem on the surface... everyone gets the same treatment.

In science, the best way for people to get to the truth is to try to prove everyone else wrong!

Now the idea of undergoing scrutiny like that in day-to-day life, is honestly an absolutely terrifying prospect. Imagine every time you put something out into the world, nobody accepts it for what it is, and EVERYONE tries to see if you're an idiot for saying it in the first place. And THEN everyone else takes your work and tries to make a better version of it.

GOOD GOD. You'd never want to try and make anything at all!

But this ain't art, folks. This is SCIENCE. Here, if someone spends their entire life trying to prove that... oh, I don't know... lemons are purple, then when a group of people inevitably rear their heads to say "Hello! We have evidence that you are wrong and that lemons are actually yellow," then that person who was trying to prove the purpleness of lemons isn't sad – not in the slightest! Because now, By Jove! thanks to the scientific method, they FINALLY know the truth about the yellow nature of lemons!

Editor's Note — Just a reminder here that the colour of lemons isn't something the scientific community concerns themselves with often. IF EVER. They probably have more important things to worry about.

There is, however, an even stranger facet. The really odd and interesting part of how the scientific method works is that, technically speaking, it's actually not possible to prove... anything.

Seriously.

In science, things are never proven to be true. Instead, they are just not yet proven to be wrong.

That's why Newton's *Theory of Gravity* is still called a theory **(the-or-ee)**, because it is still ultimately just an idea. BUT (and here's the important part) nobody has yet come up with a better idea for why things fall to the ground than the *Theory of Gravity*, and so we accept that this theory is the truth. If you want to displace it, you'd have to come up with a REALLY good alternative idea. And the last guy who did that was Einstein (with his *Theory of General Relativity*). Not that it's impossible. **It could happen... but just, you know, it was Einstein.**

So! With all that in mind, according to science, you can't prove that lemons are yellow. However, the best idea we have right now is that they probably are yellow, and so the easiest thing to do is to accept the yellowness of lemons as fact.

And that, ladies, gentlemen and everyone else is **SCIENCE!**

While you hopefully believe me when I say that I know what science is (and if you didn't before, hopefully you now understand it too) unfortunately, I still cannot confess to being a scientist myself – even with my little makeshift experiment.

I don't, however, think that is necessarily a bad thing.

While yes, this is a science book, and yes, most science books are usually written by people with actual degrees and doctorates and what not... I, personally, like to divide people who talk about science into two different groups:

Scientists and scientific communicators.

Assuming you're still the same judgemental (although hopefully slightly less so by now) person I was imagining in my head, standing in the science section of a bookshop still on the fence about whether this book is for you or not, then take just a moment to look back up at the shelf.

Editor's Note) DON'T LOOK AT THE SHELF FOR TOO LONG. THIS IS STILL THE SCIENCE BOOK YOU WANT TO BUY THE MOST.

In all likelihood, some of the books in front of you will be written by people who excel in both of these fields: doing science, and communicating it to others.

A great example of this is Professor Brian Cox. As well as being a brilliant TV host, and writer of many great science books, Brian is also... an actual scientist. You can call him Professor and everything. I mean, you *could* call me **Professor Charlie McDonnell** if you like (I won't stop you) but you'd only be doing it to make me blush a little at the thought. Unlike Brian, I don't spend my time working at the Large Hadron Collider at CERN in Geneva, trying to recreate the conditions found at the very beginning of the universe to get an insight into how it all began.

We have both been in bands, at least. But his got to no.1 in the charts in 1994. My first band's greatest achievement was getting cheered on by our parents at our school's Christmas fair (nobody else was watching).

However, the one thing that Brian and I definitely *do* share is a passion for sharing science with the world. And I also believe we share a favourite scientific communicator too:

Carl Sagan was a brilliant scientist and scientific communicator. (He is also a BIG DEAL to me personally and I will inevitably mention him an awful lot in this book.) But he actually said his great skill in communication was down to his slight inferiority when it came to the actual science part of his work. When describing himself in comparison to his peers, Carl said that he was a bit slower. He needed to break concepts down into more simple, manageable forms in order to understand them himself – and it was this that, in his own mind, made him so good at digesting and breaking down complicated ideas into bits of information that were easy for non-scientists to understand.

I know, of course, that any aspirations I have to be even 1% the scientific communicator that Carl Sagan was are ultimately a pipe dream.

BUT

I am... a bit slow.

And I do love digesting.

...SO HOPEFULLY THIS WILL BE A GOOD SCIENCE BOOK.

APPLAUSE!

(I've never been good at the hard sell.)

Here's what I can say: My goal for this book isn't to try and prove to you how smart I am, nor is it to impress you with all the science I know, or to show you that if you too had gotten your As in your science exams, then you could also be using your minimal scientific qualifications to prove to other people that you're a smarty pants who deserves their own science book.*

No.

My goal, ultimately, is to try and make you as big a science fan as I am.

That's all this is, really. It's a fan book. An unofficial annual. It's... like fan fiction. It's a poster I want you to hang on your wall.

But this ain't about *One Direction* (RIP). It's about SCIENCE.

*(Did you get that? The bit about this being a book about science?)

Being able to use your knowledge and understanding of the scientific method in order to make new discoveries, and ultimately work within the scientific community to make discoveries in the future... and to help humankind better understand its place in the universe... these aren't things that I aspire to. But they are things that I admire, and I'm here to communicate that admiration with you, by sharing what the great people of science have taught me.

And hopefully they (through me) can inspire the same love of science in you.

SO, with all this mind, let's begin...

CHAPTER 1
WELCOME TO THE UNIVERSE!

I mean, you've probably been here for a while, but who doesn't love a good welcome?

When you think about the universe, it can seem a bit foreign – which is fair enough, as most of what pops into my head when I picture the universe is stuff that I can't see in front of me. Mostly though, I think about the things that I'll never see, let alone even begin to comprehend. I mean, the universe is about 13.8 billion years old. I'm only 25 and I can barely understand myself!

Great job, Charlie. It's the first chapter and you're already comparing yourself to the entire universe.

What's more, the universe is INFINITELY large — and infinity is probably *the* most difficult concept to get your head around.

With that in mind, let's play a little game.
What is the most infinite infinity you can possibly think of?
Ok, now double that. Now double it again.
Now, sit there and double infinities for a couple of hours.
Put the book down if you like, save your arms. I can wait...

Ok, done?
Well, that infinity you just imagined?
That number of enormous scale and size that you definitely just spent
2 hours picturing because you didn't just pretend to think about it so that
you could avoid participating in my ludicrous thought experiment?

Well, unfortunately, that infinity in your head won't even bring you close
to comprehending the scale of the universe. Heck, it's so big that we don't
even know when it ends. Or if it even has an end.

However GIGANTIC the universe might seem to us
though, it's not foreign. Not in the slightest. The
universe isn't just the place that we live in, the
universe IS us. We are made of the universe!

"We are a way for the cosmos to know itself."*

Carl Sagan

So, if anything, this chapter, is really the story of **you**...

JUMP BACK IN TIME
13.8 BILLION YEARS AGO

* Sagan, Carl, *Cosmos* (Abacus, new ed. 1983)

IN THE BEGINNING, THERE WAS NOTHING.

This is obviously really hard to imagine, because we're creatures that live in time and space, and this 'nothing' I'm talking about didn't have either of those. It was the most nothingest nothing that ever existed!

This is a science book.

It was the absence of everything. It is LITERALLY incomprehensible... Look, just trust me, in the beginning, there was nothing.

...Ok, well there was this one thing: a point of infinite density, which is thought to have contained all of the mass, space and time of the entire universe. It's called the singularity. Let me just reiterate what this is for a second:

IS ANYBODY OUT THERE?

I HAVE COOKIES...

A single point of infinite density that contained everything you could possibly need to make AN ENTIRE UNIVERSE.

Doesn't make sense? Well, science doesn't yet fully understand everything, but the Big Bang Theory is still the most compelling explanation that we have for how the universe started. So, we have our nothing, and we have our singularity. Take that, add a few quantum fluctuations into the mix, and then:

It's actually a pretty common misconception that there was a giant explosion at the beginning of the universe, so don't feel bad if that's what you thought. I mean, it's called the Big Bang, after all. Who doesn't picture a giant explosion?

The truth is that the Big Bang was a very rapid expansion of matter and energy. This is a bit harder to picture, but imagine you have REALLY powerful lungs and are blowing up a balloon, so that it fills up in an instant...

Wait. It popped, didn't it? I ruined your imaginary balloon.

Go and get a new imaginary balloon out so we can do this again, but this time it's a REALLY good quality balloon, so good in fact that it will never, ever pop. It'll just keep on expanding forever! (You really are going to need good lungs.) That should give you a more accurate idea of what the Big Bang was like. (It's just a shame that the Big Expansion isn't as catchy.)

Now this expansion brought a few brand new things into existence – namely: time, space, matter and energy, but not light as it hadn't been born yet, nor would it be for thousands of years to come. It was a big ball of darkness expanding into nothing! So, not super easy to picture...

I can throw you at least one bone, though. The birth of the universe DID in fact make a noise... but it wasn't a bang. Today, sound can't travel through the vacuum of space, but for the first 760,000 years after the Big Bang, the universe was actually dense enough to carry sound waves (similar to how they function on Earth today). Using cosmic background radiation left over from the Big Bang as a guide, we've gone back in time to predict what the early universe would have sounded like... and guess what?

It sounded like it was whining.

Seriously, it sounded like a big, long whine which starts off high-pitched and then just gets lower and lower. As the universe expanded, it stretched these sound waves out, lowering their frequency and pitch.

So there we have it!
THE BIG, WHINING EXPANSION!

So, why do we call it the Big Bang? What genius came up with that name, given that it's so easy to misinterpret?

As it turns out, the man (Fred Hoyle) who came up with the term was actually a critic of the theory, and rejected it in favour of his own idea: The Steady-State theory. (Which I think sounds WAY lamer than the Big Whining Expansion.) During a radio programme in 1949 he actually coined the term Big Bang when attempting to use it to dismiss the theory, but it had quite the opposite effect. This idea of a Big Bang captured the world's imagination, and it has stuck around ever since!

ASTRONOMER FRED HOYLE

<u>How We Know it Actually Happened</u>

The Big Bang (because the Big Whining Expansion just doesn't hack it) is based on theories developed by a number of different scientists, one of the most notable being the American astronomer Edwin Hubble (1889–1953). Even if you haven't heard of the man before, you might know about the Hubble Space Telescope that was named after him, which was launched in 1990 and is still in use today!

Here's the short version: Hubble (the man, not the telescope) figured out that the universe is expanding, and used basic logic to determine that if you reverse the clock and imagine the universe contracting, it must have originated from a smaller, compact point: the singularity that I mentioned earlier.

I love how simple this idea is, which is why I wanted to bring it up first... but there's also a longer version of this discovery. In fact it's way longer, and much more complicated, so it's totally cool if you want to skip over this bit. Stick with me if you can though, because I think the long version is **SO MUCH MORE INTERESTING...**

Hubble, looking through his own telescope, observed that light from galaxies beyond our own was falling into the red part of the light spectrum, a concept which is also known as redshift.

To explain this, imagine that you are Edwin Hubble. As you look out at the universe through your telescope (which was called Hooker, no, seriously) let's pretend that you can see a galaxy in the distance that isn't moving at all. The light waves emitted from that galaxy might look something like this:

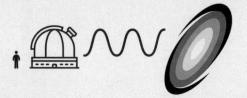

However, because this galaxy (and all other galaxies, for that matter) are moving away from us, it'd actually look more like this:

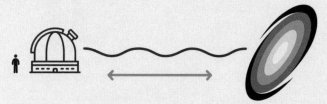

Simply put, because the galaxy is moving away from us, the light waves are being stretched out and they appear 'red.' If the galaxy was moving towards us then that light would be squished together, and would appear to be 'blue.' It was this handy little quirk of nature that allowed Hubble to figure out that the galaxies were even moving at all!

What Hubble realized was that not only was the light from all of the galaxies in the universe redshifted (meaning they were all moving away from us) but the further away the galaxy was, the more redshifted it was too! Meaning that those further galaxies were actually moving away from us FASTER than those closer to us.

Which kind of makes our galaxy, the Milky Way, seem like that awkward kid at school who just farted, and the people close to him are trying to slowly back away so that he doesn't notice that they're trying to get away from him, while the kids further away are just flat out running.

The problem with this fart-based analogy however is that it assumes that the Milky Way is at the centre of the universe, when in fact the universe doesn't have a centre at all. What that fact (that those further galaxies are moving away from us faster than those closer to us) actually shows us is that the universe is expanding uniformly. **Here's a top down, 2D view of the universe expanding to show how this works:**

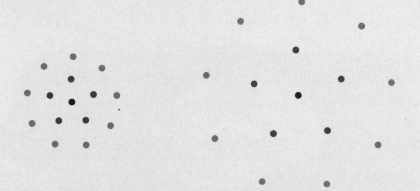

Now if you grab a ruler (because this book is *totally* interactive, kids!) and measure the distance from the centre dot to the two rings around it, and then compare the two images, then you'll find something odd. Even though the galaxies have all moved equal distances away from each other, those in the outer ring have moved twice as far away from the centre than those in the inner ring. So, from our perspective at the centre, the further away galaxies appear to be moving twice as fast... even though everything is actually expanding uniformly!

This is a good lesson to bear in mind: our perspective on the universe from our home here on Earth doesn't always show us the bigger picture. When we imagine ourselves at the centre of the universe, we are priming ourselves to make mistakes, but when we try to see our true place in the cosmos, where we are equal to everything else, only then can we begin to see the truth.

Which is actually pretty good life advice generally. Apparently this is a self-help book as well as a science book now? Let's just put that down to *GREAT VALUE!*

To recap: Hubble discovered that the universe is expanding uniformly, and realized that if you imagine that happening in reverse and turn the clock back all the way to the start of the universe, you end up with the singularity, and therefore the Big Bang.

You're So Cosmically Radiant

Now because science is science, it's not enough to say "The universe is expanding, therefore the Big Bang definitely happened!" Instead, you have to find as much evidence as you possibly can to back up your theory. Which is why now I want to talk about...

Cosmic Microwave...

Actually, let's pause for a second so I can do this:

Now that I've got the image of a microwave flying through space out of my system, let's move on to...

Cosmic Microwave Background Radiation!

The discovery of cosmic microwave background radiation (or CMB radiation, as the cool kids call it) was made in 1964 by American radio astronomers Arno Penzias and Robert Wilson. (By the way, radio astronomers investigate signals from space, they don't just point telescopes at radios.)

CMB radiation falls into the microwave part of the electromagnetic spectrum and it's found literally EVERYWHERE in the universe. Which means that it is even flowing through *you right now*! Just one cubic cm (0.39in) of space contains about 300 photons from the Big Bang. Have you ever noticed a tingling sensation in your skin? A certain heat in your belly? Sort of like you're being permeated by a 13.8 billion–year–old microwave! No? **Good**. Because, CMB radiation is impossible for us to feel. In fact, it's SO incredibly faint and hard to detect, that scientists didn't pick up on its presence until the 1960s, when their antenna equipment was sophisticated enough to notice it.

This cosmic fog is basically the earliest fossil trace of the Big Bang. As it appears absolutely *everywhere* in the universe, CMB radiation can't be coming from any one specific heat source like normal radiation does. Therefore, its complete uniformity strongly suggests that CMB radiation is what remains of the heat, or the afterglow, which was generated by the Big Bang.

Now let's get to the section that I don't think anyone could have predicted:

CHARLIE'S SUPER FUN FACTS ABOUT COSMIC BACKGROUND RADIATION!

- **The people who discovered it thought it was poop.** No joke. When Penzias and Wilson were investigating microwave emissions from the Milky Way using a large horn reflector antenna (which is about as big as a whale) they found themselves picking up a mysterious signal. Their first guess was that this interference was coming from the pigeons that were covering their antenna with droppings, so it wasn't until after they'd cleaned the antenna and stopped the pigeons from settling on it that they realized the mystery signals weren't poop after all.

- **It's almost as old as time itself.** Scientists have calculated that CMB radiation first popped up about 380,000 years after the Big Bang; because it's been around for such a long time it's a great way to learn about what the universe was like back then.

- **It's the coolest.** CMB radiation is a staggering 2.725 degrees above absolute zero (-273°C/-459°F), making it the coldest thing in the entire universe.

- **'As seen on TV!'** CMB radiation is absolutely everywhere, in fact in total it makes up 99% of all of the photons (the building blocks of light) in the universe, with the remaining 1% being found in starlight. Because it's everywhere, if you still have one of those old TVs that picks up noise when it's not tuned to any particular channel, about 1% of the static on the screen is from the Big Bang.

Before There Were Bangs

"What came first, the chicken or the egg?"

Listen, I know you were thinking about it. Are you SURE there couldn't have been anything before the Big Bang? Was this really the beginning of everything?

Here's your problem: **You're looking for the chicken.**

While most of what we know about the first few moments of the Big Bang is already quite speculative, the question of 'what happened before' is the most fuzzy of all. Like I mentioned previously, if the universe came from nothing, then the Big Bang marks the moment when time itself began. Heck, even if it wasn't the beginning of all time, for our purposes it was, as it definitely marks the start of time in our own universe.

However, there are some ideas of the universe's birth that could propose some possible answers. **(Wow, just look at how unsure that sentence is of itself.)** For example: The Cyclic Model suggests that there was another universe before our own. This universe contracted in on itself as it was dying, and it was this big crush that created the singularity – then our Big Bang happened on the other end. But then, of course, how did that previous universe come into being in the first place? With another universe before it? Surely there has to be some starting point... right?

Listen, the egg came before the chicken. Everyone should know this by now. Something that was kind of like a chicken once laid an egg, and that egg became a chicken. Maybe one day we'll find out what, if anything, existed before the Big Bang. But right now, we're an egg, and anything that came before us just isn't a chicken.

LET'S SEE IF WE CAN'T FAST-FORWARD TO THE PRESENT DAY A LITTLE FASTER...

In the beginning (again)...

As soon as the singularity occurred, an unimaginably small, hot, dense point emerged. Sort of like this:

Although, you know, way smaller than that. Smaller than an electron, so impossible to see with the naked eye. According to scientists, it was also WAY hotter than the surface of the Sun.

This point then began to expand rapidly. The first things to break apart were energy and matter, which had previously been part of the same unit, and they separated to form subatomic particles. Scientists don't refer to this as **'the universe's first breakup,'** but they should definitely start to from now on.

Then... we got lucky! There should have been an equal amount of matter and antimatter in this early universe, which would have cancelled each other out resulting in a whole lot of space with nothing in it, but there was ever so slightly more matter than there was antimatter. (Matter, by the way, is what everything we can see in the universe is made of, including us!) Without this little screw-up, the cosmos would have been pretty boring indeed.

Next up, gravity, one of the four basic forces in the universe, uncoupled itself from what scientists call the unified whole – which makes gravity the original hipster, if you will.

Soup's Up!

An infinitesimal fraction of a second later, the three other basic forces of the universe (electromagnetism, strong nuclear force and weak nuclear force) also broke off from the unified whole. (Gravity has never forgiven them for this.)

Not so unified anymore, are ya?

Another tiny fraction of a second later, due to the strong nuclear force deciding to leave, the universe experienced a cosmic inflation and grew from the tiniest point to something the size of a baseball.

The universe then continued to expand and cool, and was filled with what scientists describe as a very thin quark soup made up of hot, dense plasma. (My guess is that it must have been almost lunchtime by the time scientists got to naming this part.) Anyway, a quark is basically the smallest type of particle that you can possibly have, and they're the building blocks of everything! Plasma meanwhile is the fourth state of matter, so it goes: solid, liquid, gas, plasma. (Yes, this is the same plasma that is in plasma TVs.)

Just one second after the Big Bang, the universe cooled, enough to allow the quarks 'to get busy' and bond together to form larger particles: neutrons, protons, electrons, anti-electrons, photons and neutrinos (all words that will probably make more sense later). Several minutes passed, and then these particles started to combine to form the nuclei of the first ever elements! But just the simple ones, such as helium, lithium and hydrogen, which are the elements made up of the fewest particles. This didn't mean that these elements existed yet, but the puzzle pieces needed to make them up were ready to go.

At this point, we did have light particles (AKA the photons I just mentioned), but unfortunately these could only travel very short distances before they collided with and were absorbed by the other particles, such as electrons. (AKA the bullies of the early universe. **Let there be light, darn it!**) Due to this bullying, light wouldn't show up for another 500 million years, and so this early universe was completely shrouded in darkness...

For approximately the next 380,000 years, the universe expanded and cooled... it was a dark, silent place filled with opaque plasma soup made up of atomic nuclei and electrons.

Personally I'm now imagining *alphabet spaghetti*, but with atomic nuclei and floating electrons instead of letters. 'Elemental Spaghetti!'?! I'm sure I can't have been the first person to think of that...

Things Get Foggy

Around 380,000 years after the Big Bang, the universe had finally cooled down enough (to 3,000°C/5,432°F) for the electrons and nuclei to combine and create the first EVER atoms! It started with hydrogen, but helium would pop up later too. Finally, we were out of soup territory and the universe went from being opaque to transparent – you would have been able to see right through it! (If there was any light to see with, of course.)

In fact, this process of creating the first atoms finally DID release the photons (light particles) from their incredibly long abusive period, and they were able to travel through space, which formed our good old cosmic microwave background! However, the universe was still quite dark, because hydrogen (the scoundrel) was in a form that absorbed light. Apparently light just couldn't catch a break.

So, what do we have now? Well, we've upgraded from soup to an elemental fog, which is made up of 75% hydrogen, 25% helium (both gases) and a tiny fraction of lithium (the early universe's first metal).

What followed was a very, very long period that scientists genuinely call the Dark Ages, and this might have lasted for anything from 400 to 550 million years after the Big Bang. Objects such as stars and quasars were slowly starting to form during this time, but they weren't in any particular rush.

Basically, this was the bit where the universe just chilled out for a while and did stuff in its own time, which I totally respect and am honestly a little jealous of.

Let There Be Light Already

Nobody really knows for sure when the first stars were formed, but computer models suggest that the process of star formation probably started between 100 million and 250 million years after the Big Bang. This happened in what are called protogalaxies, which were essentially giant gas clouds that were the early versions of galaxies.

This is where things start to get pretty darn exciting (I promise). Scientists think that some of the first stars were anything from 30 to several hundred times bigger than our own Sun, and MILLIONS of times brighter – the universe only started to have light that we'd all recognize thanks to the radiation thrown off by these first stars.

Scientists also think that these first stars may have spun at surface speeds of about 1.8 million km per hour (1.1 million miles per hour) – for contrast, that's five times faster than the average speed of large stars in the Milky Way! In fact, they're sometimes called spinstars, although they're also known as Population III stars.

Come to think of it, Population III stars are actually the perfect starting point for me to talk about my...

ALL-TIME FAVOURITE SCIENTIFIC FACT (Drum roll please.)

I was honestly going to try and save this until later, but I've just gone and gotten myself excited so... screw it!

These Population III stars were basically the universe's first nuclear fusion reactors. They were SO enormous and SO hot that the immense heat and pressure found at their cores was the perfect place to start making brand new elements!

Thus far, all we had were hydrogen, helium and lithium, which are great and all that, but to use a term that I don't particularly like (but will probably resonate with the youth of today), these early atoms were pretty '**basic**.'

In order to make less basic elements, you would need the kind of heat and pressure that you can only find inside Population III stars. Because of their power, they were able to break apart these basic elements into their particle puzzle pieces, and then reform them into new, more complicated elements, such as carbon.

You see, Population III stars have a relatively short lifespan of several million years (bear in mind most stars live for about 10 billion years) and when they do die, they tend to explode in a supernova. Some scientists speculate that these supernovas might have formed black holes from which galaxies were formed, but these explosions, in conjunction with the incredibly fast spin of these stars, sent these brand new heavy elements out into the universe.

All of this means that the universe has some new, more complicated elements in it. Big deal, right? Yes. **BIG DEAL**. You see, the funny thing about atoms is that once they've formed, they stick around. If you want to break them down you need to throw them back into a big enough star or an atomic bomb (this is not to be attempted at home, or on Earth for that matter). Basically, it's not easy – so once an atom exists, it tends to just get recycled.

So, when these new, more complex atoms were sent out into the universe, they went into making more complex stars and were later recycled to make new galaxies. And planets. And the Earth. In fact, anything on this planet that contains these complicated elements, namely ALL LIFE ON EARTH, was originally formed in the hearts of stars. This includes you! You are literally made of star matter! Wait, it gets better. The different atoms that make you up were probably made in different stars! The atoms in your left hand most likely came from a different star to those from your right hand.

Wait a minute! The *Fun Science* book (or eReader) you are reading right now contains complicated elements too! Which means it is ALSO made of star stuff. Let's definitely make sure we put that on the cover!

...Skip to the End

About 9 billion years after the Big Bang, our Sun finally formed in the Milky Way galaxy – and we had sunshine! But there was no Earth for the Sun to shine on yet. Over time, some matter escaped from the Sun as it was forming, and that stuff clumped together to form the eight planets in our solar system.

The Earth was formed about 9.28 billion years after the Big Bang and many more millions of years after that, the first ever life forms appeared on Earth. Add a sprinkle of evolution and a little (well, really a LOT) more waiting and we finally get to today. That's the entire history of the universe! I told you we would get there eventually.

What We Can See

The size of the visible universe is about 92–3 billion light-years* across. Given that the universe is enormous and that light is incredibly fast, these two things work in unison to create a more comprehensible scale. For example, if we were to use km or miles instead of light years, the universe would be about 860, 981,000,000,000,000,000,000km (534,989,000,000,000,000,000,000 miles) across. So, not super easy to get your head around.

*A light year, by the way, is a unit of length in interstellar astronomy equal to the distance that light travels in one year in a vacuum.

Just to give you an idea of how fast light is, it travels at approximately 1,078 million km (670 million miles) an hour, which means it could circle the Earth almost eight times in a second. It also takes light about a second and a half to reach the Moon from Earth, and light from the Sun takes just over 8 minutes to reach Earth.

Because it takes time for light to travel to Earth, it means that when you look at the Sun, you're actually seeing what it looked like 8 minutes ago, because that light took 8 minutes to reach us. Not to try and scare you, but the Sun could actually be EXPLODING right now, and we wouldn't know anything about it until its light finally reached us... 8 minutes later.

To take this idea a step further, everything that you see around you in the world isn't actually what the world looks like right now, but it's what the world looked like a moment ago. Obviously, in day-to-day life, light is travelling fast enough that it doesn't really make any major difference to us, but the further out into the universe you look, the greater that time difference becomes. So, when you grab a telescope on a starry night and point it at our nearest galaxy, Andromeda, what you're actually seeing is what Andromeda looked like a whopping 2.5 million years ago!

So, this is basically what scientists are referring to when they talk about the visible universe, it's what we are actually seeing given the time it takes for light to travel. So, if you look back far enough, this is what you get:

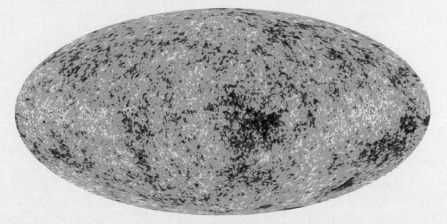

This is our cosmic horizon; it's what we see when we look SO far back, that the light is around 13.8 billion years old – which is as old as the universe itself.

That image, ladies and gentlemen, is of COSMIC MICROWAVE BACKGROUND RADIATION! The oldest thing we can possibly see. It is, as Nobel Prize-winner George Smoot referred to it,

"...like seeing the face of God."

«THE OLDEST THING WE CAN SEE IN THE UNIVERSE»

NOW THAT WE HAVE A BETTER SENSE OF HOW THE UNIVERSE BEGAN, AND HOW IT GOT TO THE POINT IT'S AT TODAY, LET'S TALK ABOUT WHAT'S ACTUALLY IN IT!

Simply put, a galaxy is a huge grouping of stars, planets, gas and dust, all held together by gravity. The word galaxy comes from the Greek word *galaxias* meaning milky, presumably because when people used to look up at the stars at night and saw the Milky Way, it reminded them of... milk. Yes, it's kind-of a weird link to make, but it's the one that we have been stuck with for thousands of years.

There are currently two leading theories which try to explain how the first galaxies formed, so it's likely that the truth involves a bit of both of them. The first theory is that galaxies formed when vast clouds of gas and dust collapsed under their own gravitational pull, and the second theory is that the young universe contained many small lumps of matter, which clumped together to form galaxies. Either way, it was probably gravity's fault.

Time to Merge

It's worth bearing in mind that the process of galaxies forming and changing is nowhere near finished; in fact our universe is evolving all the time. Smaller galaxies are often gobbled up by bigger ones, and even our own galaxy probably contains the remains of several smaller ones that it has swallowed during its lifetime. Actually it's still pretty hungry as it's currently in the process of digesting at least two small galaxies right now, and might even pull in more over the next few billion years!

The reason these mergers often happen is because, on a galactic scale, the universe is actually quite crowded. The Milky Way, for example, spans about 100,000 light years, and our nearest major galaxy, Andromeda, is about 2.5 million light years away. To us these distances obviously seem gigantic, but on a galactic scale, this actually means the distance between the two galaxies is only 25 times greater than their size.

To downsize this proportionally, the distance between me and my nearest galaxy would be a measly 12.5m (41ft). Basically, me and my mate Andromeda would definitely have to be flatmates.

Now, if Andromeda and I didn't get on too well we would easily be able to keep our distance from each other, but for galaxies, it isn't that simple. As galaxies are so massive, their gravitational force is also incredibly strong. This means that when you crowd them together, the attraction can be so strong that two galaxies latch onto each other and don't let go. Not only that, but as the galaxies get closer and closer together, that attraction increases.

So, a collision between the real Milky Way and Andromeda is pretty much inevitable. Right now they're drawing towards one another at a rate of about 400,000km (248,000 miles) per hour, and eventually they'll combine to form one, giant city of stars. It is going to take about 4 billion years for them to get around to it though, making it the most dragged out '**will they, won't they?**' love story in the history of the universe.

It'll be pretty spectacular when they do finally come together, though. When two, massive, spiral galaxies do have an encounter, huge, cold clouds of gas inside of them will be compressed, resulting in millions of new stars bursting into life all at the same time. They'll swing by each other first, and then PLUNGE into one another, forming another burst of star formations. The two galaxies will repeat this plunging action again and again, possibly taking billions of years before they merge together to form a new, elliptical galaxy... and honestly I feel like I'm writing some kind of galaxy erotic novel, so I'm going to stop now.

It's estimated that there are at least 100 billion galaxies in the observable universe, although thanks to improved telescope technology, this figure may go up to 200 billion. These galaxies range wildly in size and shape, and the smallest type that you can find is called a dwarf galaxy, which usually contains about 10 million stars. (FYI, the truly massive galaxies out there can contain up to 100 TRILLION stars, which is a thousand million.) My automatic favourite though has to be Segue 2, which is the tiniest dwarf galaxy ever found. It only has 1,000 stars! That's super adorable.

The oldest and most distant galaxy ever observed by astronomers is called **z8_GND_5296**, which frankly is just a stupid name.

Henceforth, we'll refer to **z8_GND_5296** as '**Bob**.'

Bob is about 30 billion light years away, and although he's quite small (only 1–2% the mass of the Milky Way) he is producing stars at a very rapid rate. Because astronomers are seeing Bob as he existed 700 million years ago, they hope that he'll be able to shed light on their understanding of how the earliest galaxies formed.

Types of Galaxies:

SPIRAL GALAXIES

These are the most common galaxies in our local universe. They're so common in fact that you're sitting in one right now – the Milky Way!

Astronomers believe that spiral galaxies may have formed over long periods of time by merging with smaller galaxies, and that it's this process that triggers the 'spinning' motion which results in the galaxies' typical disk shape and distinctive spiral arms. It's also thought that most spiral galaxies have a **SUPER MASSIVE** black hole in their centre (sorry, it's really hard to write the words **SUPER MASSIVE** without doing it all in bold caps) but unfortunately they're very hard to identify, even with their super-massiveness.

LENTICULAR GALAXIES

These are disk-like galaxies that basically look like their spiral cousins, but without their arms. Astronomers think that these lenticular galaxies probably used to be spiral, but are now slowly turning into elliptical galaxies – over time, spiral galaxies lose their arms as they burn through their gas and dust supplies.

ELLIPTICAL GALAXIES

These types of galaxies are distinguished by their smooth, egg-like shape, and are made up of old stars. Ellipticals don't have that much gas or dust present. They're bright at the centre, but that light tends to fade more and more towards the edges – if you look at one through a telescope, you'd be forgiven for thinking you were looking at a smudge on your lens. Also, while spiral galaxies are said to be the most common, some astronomers think it's possible that there are more elliptical galaxies than any other type! I wonder where they're hiding?

IRREGULAR GALAXIES

Now we get to the wild cards – so-called irregular as they don't really have any kind of standard configuration. These are usually smaller, and it's thought that they might form as a result of colliding galaxies or near misses.

Some irregulars are known as starburst galaxies, which is when they contain loads of gas and dust, and form young, hot stars at an exceptionally fast rate. 'Young, hot stars', making them Hollywood's favourite type of galaxy.

FUN FACT: For any *Star Wars* fans, there's also a galaxy that NASA refers to as the Death Star galaxy. The Death Star galaxy is currently blasting a neighbouring galaxy with an immense jet of radiation and particles. While both galaxies do have super massive black holes at their cores, the Death Star appears to be the more aggressive of the two. The battle continues...

Enjoy Them While You Can

Now, while I've mentioned before that the universe is expanding, what I've been keeping from you (SORRY) is that this expansion is happening at an ever-increasing rate. The more time goes on, the faster all of the galaxies in the cosmos are moving away from each other.

This is honestly a bit of a bummer.

What this means is that, eventually, all galaxies in the universe will be moving away from one another SO quickly that they'll actually reach the speed of light. If there are still astronomers around when this happens, they'll have a very different picture of the universe to the one we have today – as it means that even light itself won't be able to bridge the gap between the galaxies. When those astronomers point their telescopes to the distant heavens... all they'll see is darkness. What's more, if there are any extraterrestrials out there in galaxies beyond our own who might try to communicate with us – any signals that they might attempt to send out will never reach us.

I mean, right now, there are 100 billion galaxies in our universe that we can observe. **ONE HUNDRED BILLION.**

100,000,000,000

It's such an enormously vast and incredible cosmos out there, and the fact that it's constantly changing, means that one day there *might* be a society that looks up at the night sky and can't see anything other than their own galaxy... it makes it quite humbling to be a part of the generation that can witness the universe in its enormity.

Name:
THE MILKY WAY

Age:
13.2 BILLION YEARS OLD (GIVE OR TAKE 800 MILLION YEARS).

Anatomy:

ITS DISK MEASURES ABOUT 120,000 LIGHT YEARS ACROSS, THE BULGE IN THE CENTRE IS ABOUT 10,000 LIGHT YEARS ACROSS, AND IT'S 1,000 LIGHT YEARS THICK. (IT'S PRETTY DARN MASSIVE, BASICALLY.)

Height:
THE MILKY WAY IS ALSO BENT (OR WARPED) ALONG ITS ENTIRE LENGTH, WHICH ASTRONOMERS THINK IS DUE TO TWO OF THE MILKY WAY'S SATELLITE GALAXIES (THE LARGE AND SMALL MAGELLANIC CLOUDS) WHICH HAVE BASICALLY 'PULLED' OUR GALAXY OUT OF SHAPE. MORE GALACTIC BULLIES, APPARENTLY.

Our location:
IN THE INNER RIM OF ONE OF THE MILKY WAY'S ARMS, THE ORION ARM.

Relations:
A SUPERMASSIVE BLACK HOLE CALLED 'SAGITTARIUS A' WHICH IS IN ITS CENTRE (AND HAS A ROUGH MASS EQUIVALENT OF MORE THAN 4 MILLION SUNS PUT TOGETHER).

<<<<<THE<<MILKY<<WAY<<<<<<<<<<<<<<<<<<<<<<<<<<<<<<<<<<<<<<<<<<<<<<<<<<<<<<<<<<
<<<<<<<<<<<<<<<<<<<<<<<<<<<<<<<<<<<<<<<<<<<<<<<<<<<<<<<<<<<<<<<<<<<<<<<<<<<<<<

ULTIMATE FUN FACT: Because of the Milky Way, you're currently travelling about 1,000 times faster than a jet plane. Right now, it's rotating at a speed of 270km (168 miles) per second. Even given how fast that is, because the speed never wavers, we actually never feel how fast we're travelling. In the last hour (which I hope you've just spent sitting comfortably reading this book) you've actually travelled about 972,000km (604,000 miles) through space. This is the point where you stand up, stick your arms out, and shout **'WHOOSH!'** at the top of your lungs. I'm not joking, by the way. It's imperative to the enjoyment of this book that you do this right now. Please send a video of you doing this to me on Twitter, I'm **@coollike. I will be checking...**

The Stars of the Milky Way

Astronomers have calculated that the Milky Way contains something between 200 and 400 billion stars, and might also contain up to 100 billion planets – pretty impressive when you consider the fact that it only forms about seven new stars every year. A lot of it is populated by stars like our own Sun, but most of the stars in the galaxy are actually red dwarfs, a type of star that might be more important than we currently think. Many of the planets in the Milky Way actually orbit red dwarfs, and they emit a heat that, in theory, may make some of those planets just a wee bit more habitable for us humans. If we as a species manage to survive beyond the life cycle of our own Sun, we might have to make a new life near one of these distant red dwarfs.

The Milky Way is also home to a few supergiants, such as Betelgeuse and Rigel. (I absolutely love the names scientists have to come up with when they discover new, bigger things. 'Let's call them SUPERGIANTS!') One of the largest stars in the Milky Way is the red HYPERGIANT (very brilliant) known as VY Canis Majoris. This beast is about 250,000 to 500,000 times more luminous than our Sun.

THE GREAT ATTRACTOR

Want to hear something scary? Of course you do!

The fate of the Milky Way might not be quite as standard as we'd like to think. Out there, in the space beyond our galaxy, lies something known as the 'Great Attractor', and when I use the term 'something' I'm being as scientifically accurate as I can be. Scientists genuinely have no idea what this thing is.

What they do know, though, is that it's BIG, and it's pulling our galaxy towards it (as well as everything else around) at speeds of about 22 million km (14 million miles) per hour. Located in what's known as the Zone of Avoidance (presumably because it's a zone well worth avoiding!) this huge, unknown, invisible object is completely obscured by clouds of gas and dust in our own galaxy. Whatever it is, it's thought to have a diameter of about 300 million light years, and a mass equivalent to ten Milky Ways.

Yikes!

Although it's difficult to see, X-ray astronomy has now evolved to the point that astronomers have been able to peer into the region, and have discovered a supercluster of galaxies called Norma.

Wait, did they actually call the most threatening thing to our galaxy... Norma?

Well, actually scientists were a bit confused as to how Norma could have such a strong gravitational draw given its size. I mean, Norma's big and all that, but not quite big enough to attract other large galaxies towards her. So, it's unlikely that Norma is this Great Attractor, but is instead standing in the way of it. That sounds a bit more like a Norma to me.

FUN FACT: South African and Swedish scientists have discovered dung beetles that use the Milky Way for orientation – the first time any animal has been proven to use our galaxy to navigate. These insects use the gradient of light to dark created by the Milky Way to ensure they keep rolling their balls of poop in a straight line, and don't circle back to competitors in the dung pile!

STARS ARE ENORMOUS, LUMINOUS, INCREDIBLY HOT BALLS OF PLASMA FLOATING AROUND IN SPACE.

Scientists estimate that there are around 70 billion trillion stars in the observable universe (too many, really), the most popular of which is the Sun. You might have heard of it. It's a pretty big deal.

Truthfully, though, stars are about as unique as human beings (maybe less so in fact, as there are WAY more stars in the sky than there are people on Earth).

Types of Stars

MAIN SEQUENCE STARS

These babies make up the majority of all stars in the universe, and include the red dwarf as well as the yellow dwarf, which is the category our Sun falls into. Some of our nearest stellar neighbours, Sirius and Alpha Centauri A, are also main sequence stars.

These types of stars can vary wildly in size, mass and brightness, but they're all basically doing the same thing: converting hydrogen into helium in their cores, and releasing a tremendous amount of energy doing it. These stars can grow to more than 100 times the mass of the Sun.

RED DWARF STARS

Although we might like to think of our own Sun as being the 'average' type of star in the universe, as I've previously mentioned the most common type is the red dwarf, which makes up about three-quarters of the stars in the Milky Way.

Because they're smaller, red dwarfs are also much cooler and dimmer than our own Sun, approximately 3,230°C (5,846°F) compared to the Sun's 5,475°C (9,887°F). They also give off very little light, which makes them really hard for us to see in the night sky. They are, however, the longest living stars – because they burn dimly they also use up less energy, meaning that they have an estimated lifespan of trillions of years.

All of this, honestly, makes them a strong contender for being my favourite type of star. Yes, they're the most common, and they live the longest, but they also don't brag about it! They're happy to live out their days without worrying about anyone on Earth being able to notice them. (Sounds like a lovely life to me).

YELLOW DWARF STARS

I guess we kind-of have to do this one, don't we? This is the category that our Sun falls into; sometimes these types of stars are even referred to as 'medium' stars, and they have a typical lifespan of around 10 billion years. If you're interested, our Sun is about 4.5 billion years into its life... making it even MORE medium. Man, there really isn't anything special going on here, is there?

ORANGE DWARF STARS

There is actually something quite cool going on with orange dwarf stars – they're very interesting to scientists as, statistically, they're the most useful in our quest to finding life on other planets.

These stars stay stable for a long time, living around 15 to 30 billion years compared to the Sun's 10 billion, but they're also brighter and warmer than red dwarfs. Given this combination, it seems likely that they'd leave a larger window in which life on planets orbiting them could flourish. Not only that, but they also emit less damaging radiation than stars like our Sun, which would give DNA a better chance of emerging (which makes me even more impressed that life was able to pop up on this rock, given our chances). If we do one day find aliens out there in the cosmos, it seems most likely that they'd be orbiting an orange dwarf star.

(Should those aliens ever come to visit us, our Sun is going to be WAY brighter than theirs. Maybe they'd come out of their spaceships wearing sunglasses...)

GIANT AND SUPERGIANT STARS

RED GIANT STARS

This, in essence, is what our Sun will become after a further 5 or so billion years of its life. As a yellow dwarf dies, it begins to grow, sometimes reaching more than 400 times its original size. Basically, if humans manage to settle their differences and figure out how to combat all of the other potentially earth-shattering things that the universe can throw at us, our Sun getting all giant and red will be the final thing we will have to contend with.

RED SUPERGIANT STARS

For one of these, you have to find a star about 10 times as big as our Sun, and wait for it to reach the end of its life cycle. Only then will you end up with a **SUPERGIANT** star. These things can be thousands of times bigger than the Sun, making them the largest known stars in the universe. In cosmic terms, though, they don't tend to stick around for very long, generally burning themselves out in anything from a few hundred thousand years to possibly a few million years.

FADING STARS

BROWN DWARF STARS

Here, we have the failures – the stars that never *were*. In fact, a more accurate description of brown dwarf stars is that they are failed stars *and* failed planets. Ouch.

Size-wise, this lot sit about halfway between planets and red dwarfs, too big to be called a planet but not big enough to generate the pressure and heat needed to be categorized as a star either.

What's more, **THEY AREN'T EVEN BROWN!** They're more of a deep red colour, sometimes going into a deep pink or magenta.

Basically, they get my vote for the type of star most in need of a cuddle.

WHITE DWARF STARS

On the other end of the spectrum, white dwarf stars represent the final stage in the life cycle of a star, when it's been reduced down to nothing but its cooling, shrunken core. Eventually white dwarfs will turn into black dwarfs, which are simply cold, dead stars.

NEUTRON STARS

When a massive star collapses, usually after a supernova, this is typically what you end up with. These things are the smallest and densest stars known to exist in the universe – even with a radius of around 11km (7 miles), they can still have a mass of about 2 times that of the Sun.

B L A C K H O L E S

Ok, you got me – technically speaking, no, black holes are not stars. But hey, they're way too awesome not to mention. Besides, if you take a large enough star at the end of its life (with a core three times the size of the Sun) then, when it collapses in on itself, it forms a black hole. So they *do* come from stars... oh whatever, I shouldn't need to justify this – they're freaking **BLACK HOLES!**

These things are most definitely the strongest contender for the scariest thing in the entire universe. In terms of mass, they take the biscuit, as they are INFINITELY dense, and their gravitational pull is so powerful that, once you're near one, absolutely nothing can escape them. That's why they're black – as not even light can get away from them. Crikey, not even time and space can exist within one of these things.

The only way it's possible to observe one is by looking around them – because of their powerful gravitational fields, any nearby material is caught up and dragged in. As soon as any matter reaches what's known as the black hole's 'event horizon' then there's absolutely no turning back. Black holes simply do not give up.

One of my favourite things about black holes is that they are so powerful they distort the very fabric of space around them, and so if you find a spinning black hole then space ITSELF will start to spin around them. It's this disk of rotating space around the black hole that is known as the ergosphere. (It's easy to misread that as 'ego-sphere', which actually is a much better descriptive word for something that space itself revolves around.) If you find yourself in this area of space, it's actually impossible to stay still – as space is being dragged around, so it'll carry you along with it.

You don't even have to be near the event horizon or the ergosphere for it to be deadly. Any matter that's captured by a rotating black hole rarely just falls directly into it, but travels around the black hole before being consumed. As a result, black holes can have a lot of stuff whirring around them, with the matter nearer the black hole moving faster than the matter on the edges. This results in everything rapidly rubbing together, which generates heat. (Imagine the heat generated when you rub your hands together, but like, way hotter.) This stuff gets so hot in fact that it actually starts glowing... just like a star would! We came full circle in the end, eh?

BASICALLY
STAY AWAY FROM BLACK HOLES
THEY ARE NOT YOUR FRIENDS

FUN FACT: Stars aren't the only things that can become black holes – any amount of matter could, as long as it was shrunk down to a small enough size. If you shrunk, say, me (because why not!) and I was able to retain my mass, I'd just be an incredibly dense point, and my gravitational pull would be so strong that I'd start dragging in everything around me! (Please don't do this to me.)

OK, SO THERE'S SOMETHING I HAVEN'T TOLD YOU YET. THE UNIVERSE... HAS A **DARK SECRET.**

EDITOR'S NOTE) CHARLIE THIS ISN'T A CHAT SHOW!

Everything I've talked about so far falls under the category of 'matter'. It's what you, I, and everything we can see is made from... but this only makes up about 5% of the universe.

This is not to say that the other 95% is taken up by empty space, though – we're actually factoring space out of this equation. The rest of what makes up the universe isn't matter at all; in fact, we actually don't *really* know what the other 95% of the universe is. Our understanding here is a bit dark, hence the names: dark matter and dark energy.

Dark Matter

In the universe, there exists a material that scientists can't see because it doesn't emit sufficient light or energy to be detected. So... how do we even know that it's there in the first place? Have scientists just gotten to the point now where they think that we will believe anything they say, and so they've actually just started to make stuff up?

Well, although dark matter may appear to be completely invisible, its existence has been proven due to the gravitational effect that it has on the things around it. So although we can't see it, we can see its impact. In fact, its effect is SO impactful that, without dark matter, galaxies wouldn't even be able to hold themselves together.

As I've already mentioned, galaxies move pretty FAST. Now the faster something is in orbit, the more force you need in order to hold that thing in its orbit. So, if a star is orbiting around a galaxy incredibly fast, then it makes sense that the gravitational force at the centre of the galaxy is strong enough to hold that star in its orbit. Seems simple enough, right?

Therein lies the problem. When researchers started checking galaxies in the sky and began to calculate how much gravity a galaxy would need to hold a star in orbit, and then compared their figures with how much gravitational force the galaxy actually HAD... well, the galaxies were coming up short. They simply weren't strong enough to hold their stars in orbit. In fact, they had ten times less mass than they needed to do it! But still, there they were: galaxies in space, with stars orbiting them like nothing was wrong at all. Something invisible was helping the galaxies out, and scientists had absolutely no idea what it was.

IT WAS TIME TO SOLVE A MYSTERY!

After checking the galaxies again and again for this apparently missing matter – using infrared to look for dim stars that they might have missed, or accounting for the presence of neutron or brown dwarf stars – the scientists were still coming up short. There just still wasn't enough stuff to account for the huge gravitational force the galaxy was giving off. So, the scientists started to get desperate. Given that this mass was undetectable... they figured that the mass must be made of undetectable particles! Logic.

And it was named: dark matter. Although it also goes by a far less cool name: weakly interacting massive particles. 'Weakly interacting' as they're hard to detect as they don't interact with normal matter, and 'massive particles' because their gravitational pull is so strong, they must be massive. And, if you hadn't noticed it already, that abbreviates to 'WIMPs.' (C'mon scientists, there's no need to bully dark matter just because you haven't been able to understand it yet.)

Dark matter does more than simply hold galaxies together though – it actually dictates the structure of the entire universe.

From a distance, the universe looks a bit like a giant, messy cobweb. Galaxies are pulled together into enormous walls and thread-like shapes, criss-crossing over one another, with gigantic voids sitting between them, where very few galaxies live. Sort of like a very, very big sponge. And this – as is the way with dark matter – initially made absolutely no sense to scientists.

According to what we understood, galaxies should have been spread out evenly across space, but because they aren't, it seems as though dark matter must be to blame. See, now I'm wondering if scientists use this as an excuse for everything. 'What do you mean I left all my dirty laundry on the floor? You know that invisible dark matter has its own gravitational pull, right? I'd blame that if I were you.'

So when, exactly, are we going to be able to detect this stuff? And is it even possible to detect something that is, by its very nature, undetectable?

At the time of writing this sentence, scientists have stopped looking up at the stars and have started to try and find WIMPs at the subatomic level instead. After plummeting to the bottom of caves and tunnels to try and block out the naturally occurring particles in the universe that would just get in the way of their detectors, scientists have found... nothing. Not yet, anyway. They're also going to attempt to create dark matter in the Large Hadron Collider (AKA the enormous particle accelerator that everyone thought was going to destroy Earth, but definitely didn't).

While it's hoped that dark matter will eventually interact with normal matter so that we can see it, this may never happen. It's possible that we'll only ever be able to measure its gravitational impact – but if we don't find it soon (estimates are that we should be able to see it in the next 10 years) then there's a scarier thought to deal with. Maybe dark matter doesn't exist at all; but instead, there's something fundamentally wrong with what we know about gravity itself (sounds like a scientific mid-life crisis waiting to happen).

Dark Energy

Now it's time to talk about dark matter's even more elusive sibling: dark energy. This one is going to be really hard to explain because we know so little about it.

What we do know is how much of it there is, though. Roughly 68% of the universe is dark energy (27% is dark matter, and the final 5% is normal matter), and we know this because dark energy is affecting the rate at which the universe is expanding... but other than that scientists are, well, pretty clueless. Given that this substance makes up the majority of the universe, we should probably try and figure out what on earth it is.

It was actually via the Hubble telescope – when making observations about very distant, exploding stars – that astronomers noticed that the universe wasn't just expanding, but that it was actually doing so at an ever accelerating rate. This was a HUGE surprise, like throwing a ball up into the sky and expecting it to come back down, but instead watching it fly off into the sky, getting faster and faster as it went. The fact that it was speeding up meant there must be something out there causing it to accelerate... something completely invisible to us.

TIME TO SOLVE ANOTHER MYSTERY!

Right now, there are multiple different explanations for what dark energy is. First off, it's important to understand that space isn't *nothing*. In order for the universe to be expanding, you need to create more space, meaning that space takes up room in the cosmos just like anything else. It was Einstein who first discovered this fact.

It's possible, however, that Einstein might have been smarter than even he realized. In an old version of his *Theory of Gravity*, he predicted something he called a 'cosmological constant', the idea being that empty space could possess its own energy. He later went on to dismiss this idea, calling it the 'biggest blunder of his life.' (Poor guy!) However, his idea of a cosmological constant might be the missing piece of the puzzle in our quest to understand dark energy.

His idea was that if energy is a part of space itself, then you wouldn't lose energy as space expands... in fact, the more space is created, the more of this space-energy is created too. So, in a universe where more and more of this space-energy is created all the time, the idea of a universe expanding at a faster and faster rate would be perfectly reasonable. The only problem with this is... well, nobody understands why this cosmological constant should even be there, or why we would have just the right amount of it to cause the universe to expand.

There have been other theories of course, that dark energy might be a virtual particle, popping in and out of existence; that it could be a new kind of energy field, one that acts in a way directly opposing gravity; or maybe Einstein was just completely wrong, and gravity doesn't work how we think it does at all. **We're clutching at straws, basically – this is probably the most elusive substance in the cosmos.**

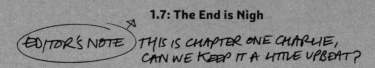

EDITOR'S NOTE THIS IS CHAPTER ONE CHARLIE, CAN WE KEEP IT A LITTLE UPBEAT?

IT'LL TAKE HUMANITY A GOOD DEAL OF CRAFTINESS, PEACE AND LUCK FOR US TO STILL BE ALIVE AT THE FINAL MOMENTS OF THE UNIVERSE (IN A COUPLE OF TRILLION YEARS, AT LEAST).

Not only would we have to figure out somewhere new to live once our planet is consumed by the Sun, but we'd probably have to live on many, many more planets after that.

But, for all of our hopeful permanency as a race, the one hurdle we may never be able to pass will be the end of the universe itself.

To be fair, scientists are still just speculating about what will happen at the end of the universe. In fact, they're not *even* entirely sure that the universe will ever end. If the universe does end though, which seems more likely than not, it'll probably happen in one of the following ways...

The Big Freeze

There ain't going to be any kind of
Netflix around for this chill! (I hate myself.)

In this scenario, the stars in the sky
will probably continue to form for
another 1 to 100 trillion years to come,
but eventually the gas needed to make new stars will simply run out. At this point,
any stars remaining will burn out what's left of them, and the universe will grow
darker and darker, eventually being dominated by nothing but black holes. These
black holes will then die out, until all that's left will be a dark, endless, empty space
– not the nicest place to raise a family.

Heat Death

Man, these names are getting more and more cheery! This idea is actually quite
closely related to the Big Freeze, although in the opposite direction.

The universe, even given its immense size, is still technically an isolated system
– and we all know what happens to things that exist inside an isolated system,
right? Yep, it's time for us to bring out everyone's favourite scientific law... the
Second Law of Thermodynamics (sexy, right?).

Just in case you aren't familiar with this law, it is used to explain how, given enough
time, anything that exists within a closed system is slowly moving towards disorder.
If you're in a closed system (which we are) then this law states that things will either
stay the same, or get worse. It's not a very optimistic law, to say the least.

So, give our universe enough time, and its chaos will increase until it reaches the
maximum level of disorder. The moment that happens, the level of heat throughout
the universe will be evenly distributed, meaning that there won't be any room for new
heat to exist and so everything will become the same temperature. Therefore, stars
will just have to... stop. There just won't be any room for any more energy to exist.

It won't stay hot for long, though. All we'll be left with is a thin soup containing some
radiation and leftover particles. If our elemental soup at the start of the universe was
chunky winter soup, this soup will be like a watery cup-a-soup. Any energy left in that
soup will disappear over time as the universe continues to expand, and we'll be left
with another dark void, with everything existing just a fraction of a degree above
absolute zero.

The Big Rip

In this apocalyptic scenario, the ever elusive and spooky dark energy is the key player in the universe's destruction. As you know already, dark energy (whatever it is) is the reason that the universe is continuing to expand. However, let's imagine a time came when dark energy became more dense and therefore more powerful than it already is, transforming into something scientists have coined 'phantom dark energy' (which sounds like a bad plot development in a sci-fi video game or film).

The idea is that this powerful dark energy would continue to stretch the galaxies further and further apart, eventually causing all matter to be pulled to pieces in its wake, effectively disintegrating everything. All the particles and radiation would be ripped apart, causing everything to shoot away from each other... for ever and ever. Here, the universe is reduced to a new singularity, where each particle becomes a loner, never being able to interact with anything ever again.

MAN that's depressing.

The Big Crunch

In the same way that dark energy could increase in power over time, it could also get weaker, meaning that the expansion of the universe could eventually grind to a stop. If it became weak enough, gravity could be the more powerful force of the two, and that'd cause the universe to begin collapsing back in on itself... resulting in a big crunch.

For a long time, this collapse of the universe would be very slow, steady, and harmless. Given enough time, though, gravity would pull everything back in together, resulting in the universe returning to the same size that it was at the Big Bang.

Out of all of the possible endings for our universe, this one is probably the most hopeful. It's unknown what would happen if the universe reached its original size, whether it would just stay like that forever... or, if some kind of expanding force kicked in, we could be looking at another Big Bang, repeating the cycle all over again!

SO, which potential death of the entire universe is your favourite? Leave me a comment below!

(Wait, that's not how books work. Sorry – Internet habits.)

Well, it kind of doesn't matter which one is your favourite anyway (although I love the Big Crunch, as it sounds like an awesome breakfast cereal) as right now, the universe doesn't seem to be showing any signs that it'll stop or slow down its expansion. Thus, scientists seem to think that the most likely ending will be the Big Freeze. Time to stock up on sweaters, I guess.

IN THE PAST, WHEN PEOPLE HAVE TRIED TO FIGURE OUT OUR POSITION IN THE UNIVERSE... WE'VE TENDED TO BE A BIT SELF-IMPORTANT.

Which is fair enough, because at first we only knew that there was the Earth, and that the Earth was everything. We were everything. The stars in the night sky were a pretty light show for us to gawk at – they were the domain of the gods above us, and the method through which they communicated with us. **We were most certainly at the centre of it all**.

Then, we learnt about the solar system... but obviously decided that the Earth must be at the centre of it. Again, we *knew* this to be true. The planets and the Sun revolved around us.

But finally, we accepted – ok, we're not the centre of the solar system... Nor are we even at the centre of our own galaxy. But our galaxy! That must be at the centre, right? If we look out at the distant galaxies, there are the same number of galaxies in all directions, and they're moving away from us! Expanding outward. That must mean we're at the centre of the universe, right?

Wrong. (But you probably guessed that already.)

In truth, the universe has no centre. Our place in it is nothing special. In fact, given the grand scale of the universe, technically, we are one of the most insignificant things in it.

"We are a speck, on a speck, on a speck, on a speck."

Neil deGrasse Tyson

However, when I think about our fragile little Earth, floating in the vast, immense cosmos... I think, hey! Don't pick on Earth! That dot is special. That dot is where LIFE lives. And I get it, it's biased of a living organism (like me) to think that life is important and special, but I don't care! Because stars don't care, galaxies don't care, but people do. Undoubtedly we're something special because, as far as we can tell, we are unique.

But all this, of course, begs the question:

Are we the only life forms in this vast universe? Are we as unique as we seem to be?

Right now, yes, we kind of are, because we're the only life forms that we know about. But the universe isn't without the possibility of life – the Kepler spacecraft has already discovered over 1,000 Earth-like planets, one of which scientists have even dubbed a 'second Earth'. There simply must be life out there somewhere, right? So why haven't we heard from anyone else yet?

The Fermi Paradox

First off, we can pretty much dismiss the idea of meeting life from outside of our local group of galaxies – they're simply too far away. If we're to believe the laws of physics (which we probably should) then nothing can travel faster than the speed of light, so it'd take aliens a long, long time to reach us.

There are about 20 billion stars similar to our Sun in the galaxy, and estimates suggest that 1 in 5 of those stars have a potentially habitable Earth-like planet orbiting them. Even if we're super conservative and say that only 0.1% of those planets have life... there would still be 1,000,000 planets with life in our galaxy. Those are some good odds!

But wait, there's more... human beings have actually started relatively late in the game – the Milky Way has been around for 13.2 billion years, but Earth has only been around for 4.5 billion. Our best estimates say that conditions in the Milky Way could have harboured life about 2 billion years into its lifespan, which leaves 6.7 billion years before Earth even existed in which life could have started! Not only that, but currently we think that an intelligent life form capable of space travel would need just a measly 2 million years to conquer the ENTIRE GALAXY! With all of that in mind, we should have seen or at least heard from ONE civilisation by now, right? So why the cold shoulder?

This is the **Fermi Paradox**, or as I like to call it:

WHY AREN'T THE ALIENS HERE YET!?... paradox.

It's possible that life at our own level is WAY trickier to get to than we think – maybe the galaxy has only just gotten to the stage where it can harbour life, or maybe the circumstances needed for life to exist are much more complicated than we can imagine. In this instance, we might just be the first ever civilization.

Or, just maybe, all life eventually finds its way to the point we're at now... but has trouble getting any further. There could be some ceiling that nobody ever gets past, e.g. war or some unseen technological catastrophe, or maybe there's just some super-intelligent race that runs around destroying any other potential life forms before they get a chance to grow. Maybe we're just too primitive for anyone to even care about us yet, or even notice us! We might just seem insignificant enough that there's no point in even bothering to make contact.

I think the scariest scenario is that we might just simply be... alone. Maybe, throughout the entire universe, there has been just one instance of intelligent life, and it's us! We're the whole thing. Yes, we'd definitely have to feel special then, but we'd also have to take that responsibility seriously. If there's nobody else around to conquer the universe, then we should probably get started!

Our Galactic Ambassadors

Even if other life forms haven't bothered or been able to make contact with us yet, we can at least be rest assured that we're reaching out to them...

First up, we have two spacecrafts currently venturing out of our solar system: Pioneer 10 and Pioneer 11! These were the first ships to ever visit the solar system's gas giants Jupiter and Saturn, and they're still on a mission to be our representatives.

Remember the chap, Carl Sagan, who said "We are a way for the cosmos to know itself" which I quoted at the start of this chapter? Well, aside from being one of the greatest scientific communicators in history, he also worked on these pioneer spacecrafts. He orchestrated the mounting of plaques on both Pioneer 10 and 11, each depicting an image of a man, a woman, and the spacecraft for scale. They also show our location in the solar system as well as in the Milky Way, and an illustration of hydrogen atoms. Should any other life forms discover these ships, they'll know who we are, where we live, and what we know.

Following in the footsteps of Pioneer 10 and 11 are Voyager 1 and Voyager 2. These two ships also studied our solar system before venturing out into the universe, although unlike the Pioneers, they still have enough power to keep transmitting radio signals until 2025! Voyager 2 also currently holds the record as the farthest manmade object from Earth; it's currently more than twice as far away as Pluto.

So, just like the Pioneer ships, Sagan and his team decided to add another bit of flair to these Voyager ships, just in case they bumped into anyone. However, instead of a plaque, these spacecraft have golden records attached to them, complete with instructions on how to play them!

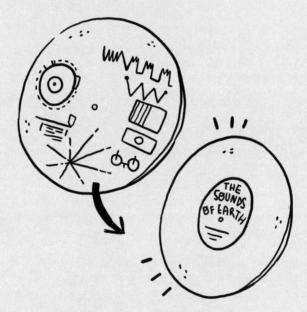

On these disks are greetings in 55 different languages, some sounds of Earth such as various animals, vehicles, weather – even the sound of a kiss, a mother with her child, and the sound of brainwaves from Ann Druyan (Carl Sagan's widow). There's a 90-minute section of music from Earth, with tracks from all around the world: Bach, Mozart, Beethoven, and other traditional songs from different cultures. They almost had The Beatles on there too with their song *Here Comes the Sun*, but their record label apparently declined due to copyright reasons! As if aliens making money from their work was a real concern to them?

There are lots of photos on it too. My favourites are probably the images that demonstrate how humans lick, eat and drink. Just imagine how weird it would be for some photo of a lady licking an ice-cream to be an alien's introduction to humanity?

To be honest, the chances of this record ever being intercepted by an alien race are so slim that you could almost consider this effort futile – but that's honestly one of my favourite things about it, the fact that we did it anyway. Out there in the universe is a time-capsule, a snapshot of our world and of humanity. It's a personal gesture, and a very human one at that, to say...

"We're here, universe! Nice to meet you!"

If we really are a way for the cosmos to know itself, what better way to get to know the universe than by saying...

"Hello?"

CHAPTER 2

CURRENTLY, YOU ARE SITTING ON THE INNER RIM OF ORION'S ARM IN THE MILKY WAY GALAXY – THE SOLAR SYSTEM.

Our solar system exists within a big, protective bubble known as the Local Bubble and within this area the solar system is currently moving through a cloud of interstellar gas, sometimes referred to as the Local Fluff.

I'm really only telling you this so that you can start adding Local Cloud and Local Fluff to the end of your mailing address and be confident in the knowledge that you're being entirely scientifically accurate when you do so.

This is the simple version of our solar system: In the centre you'll find the Sun and there are eight planets in orbit around it: **Mercury, Venus, Earth, Mars, Jupiter, Saturn, Uranus and Neptune.**

However, that picture is nowhere near complete. In our solar system you can also find a vast array of cosmic objects, including the moons of all the planets, dwarf planets (e.g. Pluto) and a vast number of other bits and bobs such as asteroids, comets, and meteoroids, which exist in regions called the asteroid belt, the Kuiper belt and the Oort cloud.

I'm a little worried that you might be imagining the solar system as being quite a manageable size, given that I've just gone on and on for an entire chapter about how ENORMOUS the universe is, but trust me – from our perspective, the solar system is still very, very big. When all is said and done, the edge of the solar system seems to be around 10 BILLION km (6 BILLION miles) from the Sun.

All of the planets in the solar system travel around the Sun in an ellipse shape – a slightly squished circle. Here on Earth it takes us about 365 days to complete one orbit around the Sun (also known as a year, although a year is different on each planet that you visit). On Mercury, the closest planet to the Sun, a year lasts just 88 days, and on Pluto, the most famous of the dwarf planets, it takes 248 Earth years for it to complete just one full orbit.

It's fun to think of the Sun as basically being the single parent of the solar system. Not only does it hold the whole dysfunctional family together but it also gave birth to the planets and moons, as well as to many of the other objects found within it. However, unlike childbirth, the story of the dawn of the solar system isn't gross in the slightest so... let's explore how it all began.

The Non-Messy Birth of the Solar System

As we currently understand it, a star and all of its planets form out of a collapsing cloud of gas and dust. Gravity, doing what gravity likes to do, pulls all the material in this cloud closer and closer together, and so the centre of the cloud gets more and more compressed. This all then begins to heat up, with that dense, hot core becoming the beginning of a new star – our Sun! Go back in time about 4.6 billion years, and you'll find a swirling, slightly glowing mess of dust and gas just like that – soon to become our new home in the universe.

As the cloud continues to compress, most of it begins to rotate in the same direction – which is actually why all the planets are still rotating in the same direction today; it never stopped spinning! This spiral cloud flattens into a disk, sort of like dough being flattened into a solar system-sized pizza base, and it's here that the planets are born... **(You want pizza for dinner tonight now, right? I know I do.)**

At this point, all of the gas and dust in this spinning disk begins to stick together and form clumps, which pick up more and more material as they orbit and steadily grow in size. Eventually, they grow so big that they become the beginnings of planets, known as planetesimals. The smaller, rocky planets then begin to form nearer the centre of the system (as the infant Sun is gobbling up all of the gas immediately around it, leaving just a little bit of rock left for those planets nearby) while the planets farther away collect the gas that the Sun can't reach, and thus they form the gas giants.

The solar system was pretty chaotic around this time – hundreds of planetesimals were fighting it out to find out who would make it as one of the final eight planets. (Now, that is reality TV I actually would watch.) Not only do these planetesimals pick up all of the material in their paths (still growing in size as they go), but they also smash into each other, they're slowly pulled together to form larger objects, and they also get sent way off-course to find new potential collisions or mergers. Basically, it's chaos.

After millions of years of this mayhem, eventually the remaining planetesimals will have cleared out most of the debris around them, and they'll form eight nice, big, lonely objects that rule over their specific regions of the solar system. Any leftover debris that didn't become one of these beasts sits in regions like the asteroid field and the Kuiper belt. The solar system, which finally looks familiar to us, has formed.

THESE DAYS IT'S COMMON KNOWLEDGE THAT THE SUN IS A STAR AT THE CENTRE OF OUR SOLAR SYSTEM, BUT SADLY, GETTING TO THE POINT WHERE THAT SIMPLE FACT HAS BEEN ACCEPTED AND ESTABLISHED HASN'T BEEN EASY...

For thousands of years, many ancient cultures viewed the Sun as a god – which seems pretty sensible to me, to be honest! The Sun is, after all, the provider of all of the energy needed for life to exist on Earth, and then some.

For the ancient Egyptians, the Sun was the god Ra. For the Aztecs it was Tonatiuh, a bloodthirsty god that required the sacrifice of human hearts! Was there genuinely not ONE day where they thought it might be worth checking whether or not the Sun would still rise and set if they didn't murder someone?

ARISTARCHUS BEING TOTALLY RADICAL

It was the Greek philosopher Anaxagoras of Clazomenae who was the first person to propose the idea that the Sun was actually a star (and so all stars were suns) back in 450 BC. The Greek astronomer and mathematician Aristarchus of Samos had the same idea, and he also suggested that the Earth revolved around the Sun. However, at the time it was common knowledge that the Sun revolved around the Earth, and so these new ideas were totally... **radical.**

Unfortunately, it wasn't until 1512 that the Polish astronomer Nicholas Copernicus brought these ideas up again – that's almost 2,000 years later! Sadly for Copernicus, his theories didn't go down very well with the Church, as at the time they were still clinging to the notion that the Earth was at the centre of everything. I mean, God wouldn't be mean enough to put us three rocks out from the middle now... would he?

The Italian philosopher Giordano Bruno tried to take things even further by suggesting that the universe was infinite, and that there might be many worlds beyond our own. However, Giordano was ultimately burnt at the stake by the Church for his blasphemous theories. Copernicus smartly waited until the end of his life before he ever published any of his ideas.

Galileo Galilei, the great Italian scientist who lived from 1564 to 1642, used his telescope to study the stars and the planets. He began to theorize that there couldn't just be one orbit for everything in the universe, and he came to the same conclusions as those before him about the structure of the solar system and the stars beyond it. Fortunately, he wasn't killed for his ideas, but he was continuously hounded by the Church, and spent the last 9 years of his life under house arrest.

It was only thanks to the efforts of 17th-century astronomers such as Johannes Kepler and Christiaan Huygens that irrefutable proof was provided; finally, it was confirmed that the Sun was at the centre of the solar system, that the planets orbited around the Sun, and that our Sun was a star. PHEW! Took them long enough, didn't it?

Name:
THE SUN

Age:
4.567 BILLION YEARS OLD. JUST
A LITTLE OLDER THAN EARTH,
WHICH IS 4.543 BILLION YEARS.

Circumference:
4,367,000KM (2,713,000 MILES),
109 TIMES BIGGER THAN EARTH.

Mass:
EQUIVALENT TO 333,060 TIMES
THAT OF EARTH'S MASS.

Volume:
1,301,019 EARTHS COULD FIT
INSIDE THE SUN.

Density:
MUCH LESS THAN EARTH'S,
BY 0.256 TIMES.

Gravity:
MUCH STRONGER THAN OURS, 28
TIMES AS POWERFUL.

<<<<<THE<<SUN<<<<<<<<<<<<<<<<<<<<<<<<<<<<<<<<<<<<<<<<<<<<<<<<<<<<<<<<<<<<<<
<<<<<<<<<<<<<<<<<<<<<<<<<<<<<<<<<<<<<<<<<<<<<<<<<<<<<<<<<<<<<<<<<<<<<<<<<<<<

Slicing Open the Sun

The Sun is essentially the engine of the solar system. It's held together and powered by gravity, and it isn't a solid or a gas, but is actually made of plasma – considered to be the fourth state of matter. Even though we might not be super-familiar with plasma here on Earth, it's actually the most common form of ordinary matter found in the universe.

The Sun accounts for almost all of the mass in the solar system – around 99% of it, in fact. The remaining 1% is left for all of the planets and everything else in the system.

The Sun is mostly made of hydrogen and contains a decent amount of helium, both of which are the two most common elements in the universe. The rest, just a minor 0.1% of its atoms, is made up of heavier elements such as oxygen and carbon. However, the Sun isn't just a simple collection of atoms; take a slice out of it (with what I imagine would be the most terrifying knife in the universe) and you'll discover the complex layers and processes going on inside. So, let's open it up!

THE CORE

While the surface temperature of the Sun might be a formidable 5,500°C (9932°F), the core comes to an absolutely whopping 15,000,000°C (27,000,000°F), making it hot enough to sustain thermonuclear fusion. And yes, this the same thermonuclear fusion used inside the hydrogen bomb – needless to say, the centre of the Sun is not a friendly place.

Because the Sun is so large, an enormous amount of mass is being pushed down onto the core by gravity, and it's this process that enables the Sun to convert that gravitational energy into light and heat. The hydrogen gases at the core of the Sun get squeezed together so tightly that they break apart, and four hydrogen nuclei combine to form one helium atom. This is thermonuclear fusion – the process that powers the Sun!

THE RADIATIVE ZONE

Right above the core, the energy is carried outwards by radiative diffusion, resulting in the very aptly named radiative zone. **(Scientists are just so gosh darn inventive with their names, aren't they?)** Unfortunately for this radiation, though, it isn't able to travel directly out of the Sun, because this zone is incredibly DENSE. There's so much plasma packed in that the radiation bounces right off it all, and travels in a kind of zigzag pattern, trying to find its way out of the Sun. Sort of like an incredibly busy party where you keep trying to leave while saying goodbye to everyone on your way out, but they keep forcing you to stick around and wanting to introduce you to new people, and are all like:

"COME ON, JUST TRY TO ENJOY YOURSELF, FOR ONCE, CHARLIE!"

...Apologies, I genuinely wasn't expecting this section on the radiative zone of the Sun to delve into my social anxiety and reluctance to 'get down' on the dancefloor – but it was just quite a useful metaphor. It'd probably be the type of party I'd avoid attending in the first place though, as it can take this radiation between 10,000 to 170,000 years to make its way from the radiative zone to the surface of the Sun.

THE CONVECTIVE ZONE

This section is just above the radiative zone, and things start to get a little cooler when you reach it – dropping from 5–10 million°K to just 2 million°K (the K which stands for Kevin, is just a measure of thermodynamic temperature similar to Celsius that scientists presumably use to make themselves feel fancy).

At this point, huge currents form, and large bubbles of hot plasma move up towards the surface of the Sun. You can kind of picture this as a boiling pot of water... but just much, much hotter. Because of this effect, the energy is transported nice and quickly through the convective zone in these bubbles, and it's allowed to pass through to the visible surface of the Sun... FREEDOM!

THE CHROMOSPHERE AND THE CORONA

These two areas make up the atmosphere of the Sun, with the lower chromosphere being quite thin – just a few thousand kilometres deep – and the upper corona reaching incredibly far, for millions of kilometres into space.

While you might expect the temperature to get cooler the further away you go from the Sun, weirdly, to begin with it actually gets hotter – it increases from around 4,300°K to 50,000°K as you pass through the chromosphere, and then temperatures in the corona can reach up to 2 million°K! Fortunately for us, here on Earth though, as the light and heat travels out from the corona it does start to cool down again, making it nice and balmy once it reaches us.

Life and Death, According to the Sun

For life here on Earth to exist, we're completely beholden to the Sun. The light and heat that it emits is what fuels all the life on this planet... and it's something we definitely take for granted. I mean, we really owe it a drink. Maybe those Aztecs who sacrificed human hearts to keep the Sun in the sky had it figured out after all! At least they were being courteous.

EDITOR'S NOTE) THE AZTECS DID NOT HAVE IT "FIGURED OUT." PLEASE REMOVE ALL WRITING THAT CONDONES HUMAN SACRIFICE.

But how exactly has the Sun enabled life on Earth? And, maybe more importantly, how could it take that life away?

THE SWEET SPOT

Luckily for us, Earth is in just the right place in the solar system for life to prosper. This area is known as the Goldilocks Zone, and like the bears' porridge in the story, the Sun's rays on the Earth are not too hot, and not too cold – they're just right. Basically, we're the right distance away – any closer and our oceans would evaporate, but if we were any further away then they'd freeze over. However, not only is Earth's location relative to the Sun perfect as a place for life to prosper, but the solar system's position in the Milky Way is also pretty convenient for the purposes of us staying alive too.

Orion's arm is probably very sought-after galactic real-estate, as it keeps us in orbit about 25,000–30,000 light years away from the centre of the galaxy. If we'd been forced to shack up nearer to the middle of the Milky Way, we'd be veering into the territory of dangerous radiation levels, comets, asteroids, and exploding supernovas.

CONGRATULATIONS, IT'S A PLANET

"But it's got great transport connections!"

Actually, speaking of the end of life as we know it...

THE DEAD SPOT

The Sun is currently about halfway through its life cycle of 10 billion years, so eventually it will swell into a red giant and then burn out completely, collapsing into a cold, white dwarf.

This is not exactly good news for us.

But wait, it gets worse! While the Sun might seem to be a steadfast, life-giving friend in our lives... it's not staying as consistent as we might like. The Sun is actually slowly heating up, as well as becoming 10% more luminous every billion years. In fact, even within just the next billion years, the heat from the Sun will be so intense that it'll boil all of the liquid water on Earth. At this point you might still find bacteria living underground, but the surface of the planet will be completely dead. Bummer.

IMMEDIATELY SURROUNDING THE SUN WE FIND THE FOUR ROCKY PLANETS, ALSO KNOWN AS THE TERRESTRIAL PLANETS: MERCURY, VENUS, EARTH AND MARS.

They're mainly composed of rocks and metals; they each have a central core that is mostly made of iron, and they all have a solid surface with familiar features to us: mountains, canyons, volcanoes and the like.

These planets are incredibly distinct in many ways, so I'm going to tackle them one by one, starting with the closest to the Sun and working my way out. I'm going to skip over Earth though... because that little guy is getting a chapter all to himself. (Lucky thing.)

Name:
MERCURY

Circumference:
15,329KM (9,524 MILES), WHICH
IS 0.06 TIMES EARTH'S.

Mass:
0.56 TIMES EARTH'S.

Volume:
0.06 TIMES EARTH'S.

Temperature:
THIS VARIES WILDLY, FROM 427°C
(800°F) DURING THE DAY TO −173°C
(−279°F) AT NIGHT.

Gravity:
0.38 TIMES THAT OF EARTH, SO
100 POUNDS ON HERE WOULD WEIGH
38 POUNDS ON MERCURY.

Rotation:
IT TAKES MERCURY 58 AND A
HALF EARTH DAYS TO MAKE ONE
ROTATION ON ITS AXIS.

Magnetic field:
VERY LOW, JUST UNDER 1% OF
EARTH'S.

Orbit:
A SUPER FAST 88 DAYS TO GO
AROUND THE SUN.

Moons:
NONE.

<<<<<P<MERCURY<<<<<<<<<<<<<<<<<<<<<<<<<<<<<<<<<<<<<<<<<<<<<<<<<<<<<<<<<<
<<<<<<<<<<<<<<<<<<<<<<<<<<<<<<<<<<<<<<<<<<<<<<<<<<<<<<<<<<<<<<<<<<<<<<<<<

Given that Mercury is the closest planet to the Sun,
you might also assume it to be the hottest in the
solar system, but that title actually belongs to Venus.
In fact, Mercury is SO close to the Sun that it doesn't
even have an atmosphere – it bears the major brunt
of the solar winds the Sun emits, which blows away
any atoms from the surface of the planet, leaving
it bare.

It's also so little (it's the smallest of the eight planets) which means its gravity is quite low, so it'd have a lot of trouble holding onto an atmosphere anyway. (Man, I don't know about you, but I'm feeling really bad for Mercury right now.)

It's because of this lack of atmosphere that Mercury can't hold its heat – while it does get very hot during the day, its temperature dips down to –173°C (–279°F) by night. This lack of atmosphere has other downsides too – where meteors usually get burnt up in the sky here on Earth, Mercury doesn't have the same protection, so it's being constantly battered by whatever falls into it. Just look at the surface of the planet and you'll see that it's covered in impact craters, just like the surface of the Moon.

Things aren't all bad for Mercury though, I promise! Because of its odd orbit and rotation, the morning Sun on Mercury rises like normal, and then it appears to change its mind and start to set again... and then rise again! This same effect happens in reverse during sunset, too.

Name:
VENUS

Circumference:
38,025KM (23,627 MILES),
VERY SIMILAR TO THAT OF
EARTH – JUST A BIT SMALLER.

Mass:
0.82 TIMES EARTH'S.

Volume:
0.857 TIMES EARTH'S.

Temperature:
462°C (864°F) ON AVERAGE.

Rotation:
IT TAKES VENUS 243 DAYS TO DO
ONE ROTATION ON ITS AXIS.

Gravity:
0.9 TIMES THAT OF EARTH'S, SO
100 POUNDS ON EARTH WEIGHS
91 POUNDS ON VENUS.

Orbit:
225 DAYS TO GO AROUND THE SUN.

Magnetic field:
VENUS DOESN'T GENERATE A
MAGNETIC FIELD.

Moons:
NONE.

<<<<<P<VENUS<<<<<<<<<<<<<<<<<<<<<<<<<<<<<<<<<<<<<<<<<<<<<<<<<<<<<<<<<<<<
<<<<<<<<<<<<<<<<<<<<<<<<<<<<<<<<<<<<<<<<<<<<<<<<<<<<<<<<<<<<<<<<<<<<<<<<

Venus is the second planet from the Sun, and is sometimes known as Earth's twin due to the two planets' similarities in terms of size and mass. In the 1930s, it was hypothesized by sci-fi writers that hidden under the thick layers of cloud on the planet's surface could lie a jungle or a swamp, warmer than Earth but probably still habitable to humans. Oh, how wrong they were...

From a distance, it's easy to romanticize Venus. Not only is it named after the Roman goddess of love and beauty, but its thick clouds reflect an awful lot of the Sun's light, making it the brightest object in the sky after the Sun and the Moon. It's so bright in fact that it's the one 'star' you can see in the sky at sunrise and sunset, and it's even bright enough to throw shadows here on Earth! It's just a shame then that, beneath those clouds, Venus is one of the most hostile places in our solar system.

To start with, it's hot – the hottest planet in the solar system, no doubt – with temperatures that average 462°C (864°F) and stay like that all day and night. The reason it's so gosh darn hot is because of the planet's thick, dense atmosphere, which is mainly made up of carbon dioxide (the greenhouse gas that causes climate change issues here on Earth) and it's this effect that traps so much of the heat from the Sun.

Beneath the surface of its poisonous sulphuric clouds you'll find an incredibly dense atmosphere with a pressure about 90 times as strong as it is here on Earth, as well as lightning and winds that reach hurricane speeds, and temperatures so hot they could melt some metals. Oh, and it's also covered in volcanoes – well over 1,000.

Basically, if you were to pay a visit to Venus, you'd be immediately choked, crushed, burnt and possibly even struck by lighting if you were particularly unlucky.

"This holiday season, visit Venus... the closest thing to hell in the solar system! Package holidays now available."

Name:
MARS

Circumference:
21,297KM (13,232 MILES).

Mass:
0.11 TIMES EARTH'S.

Volume:
15% TIMES EARTH'S.

Density:
0.714 TIMES EARTH'S.

Temperature:
–63°C (–81°F) ON AVERAGE.

Rotation:
MARS TAKES 24.5 HOURS TO COMPLETE ONE ROTATION ON ITS AXIS.

Orbit:
687 DAYS TO GO AROUND THE SUN.

Moons:
2, PHOBOS AND DEIMOS – BOTH OF WHICH HAVE BEEN DESCRIBED AS BEING POTATO SHAPED.

Gravity:
JUST 37% OF EARTH'S, SO 100 POUNDS ON EARTH WOULD WEIGH 38 POUNDS ON MARS.

Magnetic field:
VERY WEAK.

<<<<<P<MARS<<<<<<<<<<<<<<<<<<<<<<<<<<<<<<<<<<<<<<<<<<<<<<<<<<<<<<<<<
<<<<<<<<<<<<<<<<<<<<<<<<<<<<<<<<<<<<<<<<<<<<<<<<<<<<<<<<<<<<<<<<<<<<<

Mars is the fourth planet from the Sun and has two moons, Phobos and Deimos (meaning fear and panic) named after the two horses that pulled the God of War's chariot. It's funny then that, given its namesake (especially compared to Venus) Mars is currently considered humanity's best bet for a new planet to colonize – just look at the photo overleaf, taken by the Curiosity rover on the planet's surface, and you'd be forgiven for thinking you weren't just looking at a desert scene here on Earth.

Both have polar ice caps, volcanoes, canyons, four seasons and, while today the atmosphere on Mars is too thin for liquid water to exist on its surface, spacecrafts that have explored the planet have discovered clear signs that water used to flow on it. There are many channels grooved into Mars that could have only been formed by flowing water, and scientists have discovered ice on the planet, as well as some flowing water beneath its surface. They even think that 3.5 billion years ago, the climate on Mars was probably very similar to that of the early Earth – wet and warm. **If there's one place in the solar system where it seems worth it to look for life, it's probably Mars.**

EDITOR'S NOTE HANG ON – DO YOU HAVE ANYTHING NEGATIVE TO SAY ABOUT MARS CHARLIE? BECAUSE YOU'RE COMING ACROSS AS A BIT OF A FAN BOY RIGHT NOW.

Okay, let me be real with you for a moment... I might have a bit of a bias towards Mars. It's not exactly THE fanciest planet in the solar system (I'd probably give that title to Saturn) but I do have a bit of an attachment towards it, probably because of one day back in July 2010.

Did I ever tell you about the time I visited NASA?

When I arrived at JPL (the NASA Jet Propulsion Laboratory in California) I was greeted at security by D.J., one of the smartest people I've ever met. The visit featured many standout moments for me: seeing the first ever digital image from the surface of Mars, which had to be coloured in by scientists here on Earth in a very scientific paint-by-numbers way; looking at the life-sized replicas of the Voyager spacecraft, and then suddenly realizing that this was the place where those intergalactic ambassadors were actually made... but hearing about history can never really compare to seeing it happen before your eyes.

As we passed by JPL's clean room, D.J. took me up to a window and allowed me to watch as the Curiosity rover – the one that's currently on Mars – was being built.

Now at the time I was just like: "Neat! That rover is going to be on Mars some day!" But it's only with hindsight that I'm able to fully appreciate how cool this moment actually was. Today, I can't help but feel a rush of awe every time I see a photo taken by Curiosity from the surface of Mars, or even just hear a silly story about how it has sung itself 'Happy Birthday'. All because D.J. at NASA saw some of my videos on the internet, and sent me an email asking if I'd like a tour of where he works.

So, there you go. That's why I'm biased towards Mars: because I once met someone who lives there, alright?

BEYOND THE TERRESTRIALS AND THE ASTEROID BELT LIE THE MOST MASSIVE PLANETS IN OUR SOLAR SYSTEM — THE FOUR GAS GIANTS: JUPITER, SATURN, URANUS AND NEPTUNE.

Unlike the four rocky terrestrial planets (which seem more like planetary pebbles in comparison) these beasts are made up mostly of gas, and as such have no solid surface on which you could stand. They're also much further away and more spread out than the terrestrials are; in fact Neptune and Uranus are so far out and icy cold that they're actually classified as ice giants*.

Continuing on our journey outwards from the Sun, the first of the giants that we meet is...

---

* Ice giants aren't called that because they're made of ice – in fact, they're actually mostly made of heavy elements, like oxygen, carbon, nitrogen and sulphur.

Name:
JUPITER

Circumference:
439,264KM (272,946 MILES),
11 TIMES THE SIZE OF EARTH.

Mass:
318 TIMES THAT OF EARTH.

Volume:
1,321 EARTHS COULD FIT INSIDE
JUPITER.

Density
0.24 TIMES EARTH'S.

Temperature:
ABOUT −148°C (−234°F) AT THE TOP
OF ITS CLOUDS.

Gravity:
2.64 TIMES EARTH'S, SO 100
POUNDS ON EARTH WOULD WEIGH
253 POUNDS ON JUPITER.

Magnetic Field:
20,000 TIMES EARTH'S.

Moons:
67 AS OF 2015 – JUPITER LOSES
AND GAINS THEM, SO THE COUNT
GOES UP AND DOWN. THE FOUR
LARGEST MOONS ARE: IO, EUROPA,
GANYMEDE AND CALLISTO.

Rotation:
JUPITER SPINS ONCE ON
ITS AXIS EVERY 9 HOURS AND
55 MINUTES.

Orbit:
IT TAKES JUPITER 11.86 YEARS
TO COMPLETE ONE TRIP AROUND
THE SUN.

Rings:
IT HAS THREE RINGS, ALTHOUGH
THEY'RE VERY HARD TO SEE.

<<<<<<P<JUPITER<<<<<<<<<<<<<<<<<<<<<<<<<<<<<<<<<<<<<<<<<<<<<<<<<<<<<<<<<<<<<<<<<<<<<<<<<<<<<<<<<<<<<<<<<<<<<<<<<<<<<<<<<<<<<<<<<<<<<<<<<

Jupiter is the fifth planet from the Sun, and the biggest in the entire solar system. It alone makes up two-thirds of the mass of all of the planets combined and, by volume, it'd be possible to fit all of the other planets in the solar system inside of it.

It also acts as a bodyguard, as it does a good job of keeping Earth protected from a lot of the dangerous junk that floats around in the solar system. Because of the planet's immense size it has a very strong gravitational pull, and if you combine that with its incredibly powerful magnetic field then you've essentially got a giant magnet, pulling in all of those asteroids and comets that might otherwise end up on a collision course with Earth.

We're lucky to have Jupiter on our side however, as it's possible it might have been a bit of a bully during the early days of the solar system. It's thought by scientists that, around 4 billion years ago, there actually used to be not four, but five gas giants, and astrophysicists think it's most likely that it was Jupiter's gravity that forced out that missing planet! With any luck, that fifth giant might have found another star to orbit around by now... but more likely, it's probably just roaming around the Milky Way alone.

Just like Venus, Jupiter wouldn't exactly be a nice place to visit. Its pressure is so powerful that it actually squishes gases into liquid, meaning that any spacecraft (or person) attempting to travel through its colourful clouds would be promptly crushed and then melted.

Probably Jupiter's most famous feature though is its great red spot – a giant, high pressure storm, sort of like an enormous hurricane. It's so large in fact, that you could fit three Earths inside it. What's more, this storm has been raging for possibly 350 years or more!

Name:
SATURN

Circumference:
365,882KM (227,348 MILES),
NINE TIMES LARGER THAN EARTH.

Mass:
95 TIMES EARTH'S.

Volume:
AROUND 755-764 EARTHS COULD FIT
INSIDE SATURN.

Density
0.125 TIMES EARTH'S.

Temperature:
-178°C (-288°F) AT THE TOP OF ITS
CLOUDS.

Rotation:
SATURN SPINS ONCE ON ITS AXIS
EVERY 10.7 HOURS.

Orbit:
IT TAKES SATURN 29.5 YEARS TO
TRAVEL ONCE AROUND THE SUN.

Gravity:
VERY SIMILAR TO EARTH, 100
POUNDS ON EARTH WOULD WEIGH
107 POUNDS ON SATURN.

Magnetic field:
578 TIMES MORE POWERFUL
THAN EARTH.

Moons:
62, WITH 53 OF THOSE KNOWN
AND 9 STILL "AWAITING
CONFIRMATION". THE MOST WELL
KNOWN ARE: TITAN, ENCELADUS,
LAPETUS, MIMAS, TETHYS, DIONE,
AND RHEA.

<<<<<<P<SATURN<<<<<<<<<<<<<<<<<<<<<<<<<<<<<<<<<<<<<<<<<<<<<<<<<<<<<<<<
<<<<<<<<<<<<<<<<<<<<<<<<<<<<<<<<<<<<<<<<<<<<<<<<<<<<<<<<<<<<<<<<<<<<<<

Saturn is the sixth planet from the Sun, and the second
largest in the solar system after Jupiter. It's popularly
known as the 'Jewel of the Solar System' for, well...
just being super pretty. (It does have some damn
fine rings.)

Get up close to Saturn and you'll discover that its rings mainly consist of ice, with some rocks and dust in there too. These rings have probably been there since the planet first formed, and they've accumulated in a number of different ways – from passing comets, meteorite impacts, and from the planet's gravity simply pulling in material from its many moons.

Because Saturn's rings are made almost entirely of ice, when light from the Sun hits the rings, it reflects that light right back. So, the reason we can see Saturn's rings so clearly here on Earth is because they are literally shiny!

Looking beyond those rings for a moment (as I'm sure Saturn would like us to – inner beauty is important as well, you know) the planet has some really interesting phenomena going on. Like Jupiter, Saturn is a very windy planet, and it sports a feature that, for my money, is even more interesting than Jupiter's great red spot: the hexagon. This is a six-sided jet stream with a massive, rotating storm at its centre, that sits right on the top of Saturn's north pole, spanning about 32,000km (20,000 miles). It's also in the shape of a hexagon. A HEXAGON! In space! I mean, just look at this thing:

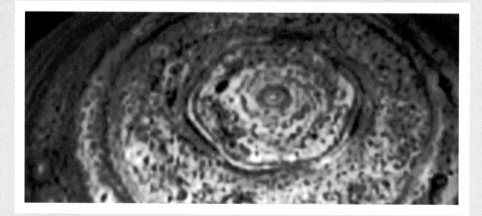

Then, of course, there's the matter of the planet's density... it's the only planet in the solar system that is less dense than water – meaning that, if you were able to get your hands on a big enough swimming pool to put it in, Saturn would float! Like a planet-sized rubber ducky.

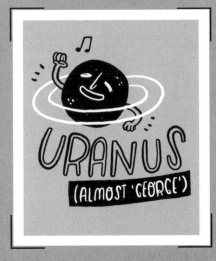

Name:
URANUS

Circumference:
159,354KM (99,018 MILES),
NEARLY FOUR TIMES
THE SIZE OF EARTH.

Mass:
ABOUT THE SAME AS 14.5 EARTHS.

Volume:
63 EARTHS COULD FIT INSIDE
OF URANUS.

Density:
0.23 TIMES EARTH'S.

Gravity:
SIMILAR TO EARTH, 100 POUNDS
ON EARTH WOULD WEIGH 91 POUNDS
ON URANUS.

Temperature:
ABOUT −220°C (−364°F) AT THE
TOP OF ITS CLOUDS.

Magnetic field:
48 TIMES MORE POWERFUL
THAN EARTH.

Rotation:
URANUS TAKES 17 HOURS TO
COMPLETE ONE ROTATION ON
ITS AXIS.

Rings:
13, BUT THEY'RE VERY FAINT
AND CAN ONLY BE SEEN WITH
SPECIAL EQUIPMENT.

Moons:
27 KNOWN MOONS. NONE ARE VERY
BIG, THE LARGEST OF WHICH ARE
OBERON AND TITANIA — ALMOST
ALL OF THEM ARE NAMED AFTER
SHAKESPEARE'S PLAYS.

Orbit:
IT TAKES URANUS 84 YEARS TO
TRAVEL ONCE AROUND THE SUN.

<<<<<<P<URANUS<<<<<<<<<<<<<<<<<<<<<<<<<<<<<<<<<<<<<<<<<<<<<<<<<<<<<
<<<<<<<<<<<<<<<<<<<<<<<<<<<<<<<<<<<<<<<<<<<<<<<<<<<<<<<<<<<<<<<<<<<

Firstly, it's pronounced

YOOR-un-nus — *(there's no "A" sound in there.)* Although, honestly, it does
still sound pretty close to what you were thinking. Uranus is the seventh
planet from the Sun, the third largest in the solar system, and it almost ended
up with an entirely different name: **George.** Seriously.

Uranus was the first planet ever to be discovered in the modern age – most of the planets are clearly visible in the night sky going as far out as Saturn, and so they've been known about for millennia. Uranus, however, was deemed a planet by Sir William Herschel (it was previously thought to be a star) and he decided to call it George's Star, after the king at the time, George III. Unfortunately, for George, the astronomical community overruled Herschel and decided upon Uranus instead, much to the pleasure of school children across the planet.

Just like its neighbour Neptune, Uranus was reclassified in the 1990s as an ice giant, partly due to its very cold temperature. It's not made of ice, though – it is still a gaseous planet – it's simply the methane in its atmosphere that turns it that blue colour.

Probably the most unique thing about Uranus though is its rotation – it's tilted on its axis by a massive 98 degrees, meaning that instead of spinning around like all of the other planets do, it rolls around the Sun like a ball! It's thought this probably happened early on in its life when it collided with another planetary body, sending it a bit off-course. It also rotates in the opposite direction to the other planets (with the exception of Venus).

All of this makes Uranus so eccentric that it's almost like the planet is doing some kind of silly dance to try and distract us all from its funny name. (It's not working, Uranus.)

Name:
NEPTUNE

Circumference:
154,705KM (96,129 MILES), NEARLY FOUR TIMES AS BIG AS EARTH.

Mass:
17 TIMES THAT OF EARTH'S.

Volume:
58 EARTHS COULD FIT INSIDE OF NEPTUNE.

Density
0.3 TIMES THAT OF EARTH'S.

Temperature:
−218°C (−360°F) AT THE TOP OF ITS CLOUDS.

Rotation
IT TAKES NEPTUNE 16 HOURS TO SPIN ONCE ON ITS AXIS.

Orbit:
NEPTUNE TAKES JUST OVER 165 YEARS TO TRAVEL ONCE AROUND THE SUN.

Density:
0.6 TIMES EARTH'S.

Gravity:
QUITE SIMILAR TO EARTH'S, 100 POUNDS HERE WOULD WEIGH 114 POUNDS ON NEPTUNE.

Magnetic field:
27 TIMES MORE POWERFUL THAN EARTH'S.

Moons:
13 CONFIRMED, AND ONE IS PENDING. ITS MOST SIGNIFICANT MOON IS TRITON.

```
<<<<<P<NEPTUNE<<<<<<<<<<<<<<<<<<<<<<<<<<<<<<<<<<<<<<<<<<<<<<<<<<<<<<<<<<
<<<<<<<<<<<<<<<<<<<<<<<<<<<<<<<<<<<<<<<<<<<<<<<<<<<<<<<<<<<<<<<<<<<<<<<<
```

Neptune is the eighth and therefore most distant planet from the Sun. Just like Uranus, it's the methane in its atmosphere that turns it that blue colour, and it's technically classified as an ice giant.

Sadly for Neptune, it simply... isn't a very interesting planet. That's my opinion, anyway. Maybe it's just that it's far enough away and was discovered recently enough that not much cool stuff is known about it yet – it's only been visited once after all, by the Voyager 2 spacecraft, and it only did a quick fly-by.

But there is still some hope for Neptune, as in the future it might even be able to rival the beauty of Saturn. Its largest moon, Triton, has a strange backwards orbit that means, year by year, it's getting closer and closer to the planet. This probably means that Triton used to actually be a dwarf planet that Neptune captured in its orbit. When these two bodies finally collide, Triton will be torn to pieces and will form a ring system around Neptune, just like Saturn's! This won't be happening for about 3.6 billion years, but when it does, I promise to rewrite this section and tell everyone that it's my new favourite planet. (Or, assuming I am dead by then, can someone else do it for me, please?)

WHEN I WAS TALKING ABOUT URANUS AND NEPTUNE EARLIER, I CASUALLY MENTIONED THE FACT THAT NEITHER OF THEM ARE CONSIDERED TO BE **GAS GIANTS** ANYMORE, BUT WERE ACTUALLY RECLASSIFIED IN THE 1990S TO BE **ICE GIANTS** INSTEAD.

And, if I'm to take a guess... I'll say you probably weren't exactly 'outraged' by that reclassification?

That's right, folks! It's time to tackle **Plutogate**. (I don't think anyone ever actually called it that, by the way. I'm just trying to add some drama.)

The formal definition of a planet is as follows:

A planet is a celestial body that **(a)** is in orbit around the Sun, **(b)** has sufficient mass for its self-gravity to overcome rigid body forces so that it assumes a hydrostatic equilibrium shape...

Actually, lets just stop there for a moment – according to everything stated so far, Pluto would be classified as a planet: it's in space, it orbits around the Sun, and it's round. All good! However:

...and **(c)** has cleared the neighbourhood around its orbit.

Oh, right...

Pluto, unfortunately, never did this. Instead, it lives in the busy Kuiper Belt just beyond Neptune, surrounded by objects of a similar size, and joins a new classification along with Ceres, Eris, Haumea and Makemake: the Dwarf Planets.

For some, this apparent demotion for Pluto (which was previously considered a planet until 2006) was a cause for much sadness and chagrin. But if we're going to personify Pluto, just note this for a second: as a planet, Pluto was the smallest, coldest, slowest and most irregular one of the lot. However, as a dwarf planet, not only is it a member of its own new group, on equal footing with the rest, but it was also the first member of that group to ever be discovered. No longer 9th, but 1st.

I am, however, a child of the nineties. So, unlike the other dwarf planets, I am going to give it a little section all to itself.

Name:
PLUTO

Circumference:
7,232KM (4,493 MILES), WHICH IS
0.18 TIMES THAT OF EARTH.

Mass:
0.002 TIMES EARTH'S.

Volume:
0.006 TIMES EARTH'S.

Density
0.37 TIMES EARTH'S.

Temperature:
AN AVERAGE OF –230°C (–382°F).

Rotation:
IT TAKES PLUTO 6.4 DAYS TO MAKE
ONE ROTATION ON ITS AXIS.

Orbit:
PLUTO TAKES 248 YEARS TO
TRAVEL ONCE AROUND THE SUN.

Moons:
FIVE: CHARON, KERBEROS, STYX,
NIX AND HYDRA.

Gravity:
100 POUNDS ON EARTH WOULD ONLY
WEIGH 7 POUNDS ON PLUTO.

Magnetic field:
VERY LOW, ABOUT 1% THAT
OF EARTH'S.

<<<<<<P<PLUTO<<<<<<<<<<<<<<<<<<<<<<<<<<<<<<<<<<<<<<<<<<<<<<<<<<<<<<<<<<<
<<<<<<<<<<<<<<<<<<<<<<<<<<<<<<<<<<<<<<<<<<<<<<<<<<<<<<<<<<<<<<<<<<<<<<<<

Pluto was first discovered in 1930 and, although it does
bear the name of the god of the underworld, Pluto is,
frankly, adorable. It's just a little bit smaller than our
Moon, and in 2015 when the New Horizons spacecraft
paid it a visit, it revealed a variety of surface features
including the Tombaugh Regio – a smooth region of
Pluto that looks, undeniably, like a giant heart.

Pluto also has a delightfully eccentric orbit – so odd, in fact, that for 20 years out of its 248-year-long journey around the Sun, Pluto actually moves inside the orbit of Neptune. So, from 1979 to 1999, not only was Pluto still considered a planet, but it was actually the 8th planet from the Sun, not the 9th.

So, yes, Pluto may no longer be a planet, but just because its not big enough to clear the neighbourhood around its orbit... well, that shouldn't make it any less worth your time.

Space Junk

Here's a handy-dandy guide to the definitions of the different bits of space debris... (with some puns that may or may not help you to remember them).

- **Asteroid:** A relatively small – compared to everything else in the solar system – inactive, rocky body orbiting around the Sun. (It can't be assed-eroid to be active...)
- **Comet:** A relatively small, at times active object, whose ices can vaporize in sunlight, sometimes creating a tail of gas and/or dust. (It's comet-ted to staying active!)
- **Meteoroid:** A small particle from a comet or asteroid in orbit around the Sun. (Looks like this little meteoroid is trying to mete-avoid colliding with Earth!)
- **Meteor:** The light phenomena which occur when meteoroids enter Earth's atmosphere and vaporize. (It's burning up in the sky! What a treat-eor.)
- **Meteorite:** A meteoroid that survives its passage through the atmosphere and lands on the surface of Earth. (It's survived the trip to Earth – it's mete-alright!)

EDITOR'S NOTE ⟩ NOT SURE IF PUNS ARE SUITABLE AS MEMORY AIDS. DEFINITELY SURE THAT THEY AREN'T FUNNY.

Every single day, Earth is bombarded with more than 44 metric tons of meteoric material – fortunately for us however, thanks to Earth's atmosphere, most of this stuff burns up before it can ever reach the surface. Even with Earth's handy protective cloak, though, it doesn't shield us from everything that could come our way.

- Once a year, an asteroid the size of a car will fall into our atmosphere, creating a fireball that will dissipate before it hits the ground. (Phew!)

- Every 2,000 years, a meteoroid the size of a football field will hit the Earth, causing significant damage to the area around it (Oh...!).

- Every few million years, Earth is visited by an object large enough to destroy an entire civilization. (Darn!)

- Any body larger than 1 or 2km (1 mile) that reaches us could cause effects worldwide, and right now the largest known, potentially hazardous asteroid to us is Toutatis, which is around 5km (3 miles) across. (Well... heck!)

Well over 1,000 asteroids have been classified as posing a threat to Earth... no wonder people in ancient times used to see this stuff as being signs of misfortune and tragedy – sounds like a good hunch to me! However, how serious a threat are they, really?

Well, as long as scientists can figure out if an asteroid is on a collision course with Earth 30–40 years before it's due to hit us, then we should have time to react to it... although obviously the technology required to combat it would have to be developed in that time. I wonder why scientists don't use THAT as an argument when they're asking for more funding.

ASTEROIDS

The ones that couldn't be

Assed-eroid

– to be active, remember?
(Of course you do – you can't beat
a good pun-based memory aid.)

These things are basically rocky, airless little worlds, too small to ever be considered dwarf planets, instead having to settle for the term 'minor planets'. You can find tens of thousands of these gathered in the asteroid belt, which is a vast ring sitting between the orbits of Mars and Jupiter, separating the terrestrial planets from the gas and ice giants. Thanks to Jupiter's massive gravitational force (and a few close encounters with Mars) these asteroids can get thrown out of their belt at a moment's notice, and fly out into space... although hopefully not in our direction.

Asteroids range in size – the largest is Vesta at around 525km (326 miles) across, and one of the smallest is 1991 BA, which is only about 6m (20ft) wide. Most of them are irregularly shaped, and they're diverse, too – while more than 150 asteroids are known to have small companion moons, some of them seem to have gone on to become moons themselves, with Mars' moons Phobos and Deimos being very likely candidates. And not everything in the asteroid belt is actually an asteroid – it's also the home to some comets, as well as the dwarf planet Ceres.

Unlike comets, the committee that names asteroids doesn't seem to be particularly strict, resulting in some asteroids that definitely don't sound like they'd be particularly threatening to Earth. **For example, there's one named after the musician Frank Zappa, and another called Mr. Spock, which is named after a cat, who in turn was named after the character from 'Star Trek'. Sadly for me, though, my cat Gideon will never have an asteroid named after him as the International Astronomical Union now discourages naming asteroids after pets, so Mr. Spock is going to remain unique forever.**

COMETS

Comets are generally thought to come from two different areas of space: The Kuiper belt and the Oort cloud. These are two busy regions beyond the orbit of Neptune, with the Oort cloud being the further of the two. In fact, the Oort cloud is so far away that... it hasn't actually been discovered yet! While it remains just a theory, it very likely does exist as it's the only way we can explain the existence of

some comets with incredibly long orbits that
have been observed. (If there's one place
in the solar system those comets could have
come from, it oort to be the Oort cloud!)

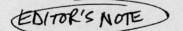

EDITOR'S NOTE

THERE ARE ONLY SO MANY TERRIBLE PUNS I CAN BE BOTHERED TO REMOVE, CHARLIE.

Unlike the inactive asteroids (which are essentially just huge, floating rocks) comets are like space snowballs which orbit around the Sun. They're made up mostly of ice, as well as frozen gases, rocks and dust, and when one of these things gets close enough to the Sun it starts spewing out all of its frozen matter, forming a sort of glowing halo around the comet. This halo can get stretched out to form a tail, and some comet tails can go on for millions of kilometres!

People weren't always quite so scientific about comets, though. Throughout history, different cultures around the world have seen them as clear symbols of war and death, earning them titles such as 'harbinger of doom' and 'menace of the universe'. As presumptuous as these assertions are (I mean, just get to know a comet before you make claims like that!), it is possible to understand where these ideas might have come from.

The sky at night fits a constant, recognizable pattern – everything in it, the stars and the planets, can be measured and predicted.

So, what happens when a comet suddenly rears its head, appearing as if from nowhere? With a huge, glowing tail in its wake and everything? It seems fair enough that people would be a bit perturbed by its presence. In fact, even as recently as 1910, businesspeople tried making the most of the public's comet fears by selling items such as gas masks and comet pills to help 'protect' them when the Earth passed through the tail of Halley's Comet!

The reality about comets however is that they actually are very predictable – it's just that their orbits are incredibly large, so we don't see them here on Earth very frequently. For example, Halley's Comet only stops by to say hello every 75–76 years (it'll be back in 2061) and some long orbit comets can only be seen every 200 to 1,000 years! No wonder they came as a surprise...

NOW, I'VE DONE MY ABSOLUTE BEST TO FILL THIS CHAPTER WITH AS MANY FUN AND INTERESTING FACTS ABOUT THE SOLAR SYSTEM AS I POSSIBLY COULD — AND BELIEVE ME, THERE'S A LOT OF SUPER COOL STUFF I HAD TO MISS OUT.

(Hey, maybe I could fit all that into another book? Wink-wink, nudge-nudge?) However, up until right now, I've dutifully managed to hold on to my all-time favourite fact about the solar system.

There is a slight problem, though... and it's that this fact is undoubtedly one that you're going to know already. I'm not about to drop another **'We are all made of stardust!'** bombshell on you – instead I just want to spell out one simple truth, which is:

The solar system... is all around us!

(Man, this part is going to be harder than I thought.) Let me cut to the chase – if there's one message I'd really love you to take away from this section of the book, it's this:

LOOK THROUGH A GOSH DARN TELESCOPE.

Please! Buy one, or find someone who has one, or just make one yourself out of a bunch of sellotape and toilet roll if you have to because, believe me, there's nothing quite like seeing a planet for real.

EDITOR'S NOTE FUN SCIENCE DOES NOT TAKE RESPONSIBILITY FOR ANY NON-FUNCTIONING TOILET ROLL TELESCOPES.

This is really what I mean when I say **'the solar system is all around us.'**
I mean that it's RIGHT THERE. Most importantly, it's a REAL THING!

You already know intellectually that the planets are out there in the distance – I mean, you've seen the photos! You might have even seen them in the night sky with your naked eye. But I promise, when you see a planet through a telescope, it becomes real in a way that all the science books and photos in the world will never quite help you to grasp. I do think that having as much information about the planets in advance will help you appreciate them more when you do see them but, man...

I can still remember it so clearly – that feeling I got the first time I saw Jupiter in all its glory. Thinking... it's a ball of GAS! In the sky! It's like Earth – a real, physical place! And don't even get me started on the first time I ever saw Saturn, I mean... those RINGS! They're so beautiful and... REAL! Actual rings around a planet – MILLIONS OF BALLS OF ICE just stuck spinning around that giant mass of gas. **They're actually... THERE!** I know, I know... as much as I want to try, I doubt I'll ever be able to accurately articulate this feeling of mine. NO MATTER HOW MUCH ALL-CAPS I USE. And to be honest, if you do go out now and look through a telescope for the first time, there's no way I can say for sure that your experience will match up with mine. Maybe you'll just think I'm some loon who seems to really love gas orbs (which, to be honest, I probably am). But, I had to try – just on the off-chance that you'll decide to peep down the lens of a telescope yourself and have the same experience that I did.

Sorry – I just sort of... poured my heart out a bit there, didn't I? But I guess that's what being a science fan is all about – allowing yourself to get unapologetically excited and moved by the real world.

Frankly, that doesn't seem so crazy to me.

CHAPTER 3

THE EARTH

THE LARGEST OF THE ROCKY TERRESTRIAL PLANETS

THE THIRD PLANET FROM THE SUN

THE ONLY PLACE IN THE ENTIRE UNIVERSE WHERE LIFE IS KNOWN TO EXIST (SO FAR).

EARTH
Pale Blue Dot.
Taken on February 14, 1990, by
the Voyager 1 space probe from a
record distance of about 6 billion km
(3.7 billion miles).

WHILE OUR PLANET IS 4.543 BILLION YEARS OLD, FOR ABOUT 4 BILLION YEARS OF THAT TIME, LIFE AS WE KNOW IT DIDN'T EXIST. IN FACT, OUR OWN SPECIES HAS ONLY BEEN AROUND FOR ABOUT 200,000 YEARS, MEANING THAT HUMANITY HAS COLLECTIVELY ONLY SEEN ABOUT 0.004% OF EARTH'S HISTORY.

Basically, we've missed a LOT.

Instead of us jumping into a time machine to observe for ourselves what early Earth might have looked like – an idea that might not be quite as farfetched as you might think, (*see* Chapter 10 for more on that) – currently, our best tool for figuring out what the Earth used to look like is...

Rocks!

Ok, maybe it's not quite as exciting as actual time travel... but still! We all know what people say about rocks...

They... TOTALLY ROCK!

Let's get real for a moment: We're about to jump (yes, you and me) from the awe-inspiring worlds of astronomy and cosmology – the studies of some of the most breathtaking wonders that the universe has to offer – to geology, also known as... the study of rocks. Allegedly, some of the books that have been written on this subject are the dullest works ever published.

HOWEVER, what I'm not trying to do here is give you some kind of **dull warning**; far from it, in fact. While I will definitely bring up words such as sedimentary and radiometric and there might even be mention of a certain Precambrian eon, I have absolutely no intention of boring you with my rock-talk. The Earth is a genuinely fascinating place, and so in this chapter I've waded through all the dull stuff in order to present you with the creamiest of the geological crop that I could manage.

This is Fun Science, after all.

With all that said, let's begin our journey into the history of the Earth with the incredibly exciting question...

How Old is That Rock?

Stay with me, guys.

You can think of geologists sort of like... rock detectives. However, while most detectives will only work on cases that have happened in recent history, geologists are trying to piece together a story that started billions of years ago. So, really, they're more like...
Super Rock detectives!

Now, where the normal detective's key technique for solving crimes is poking around for fingerprints with a giant magnifying glass (I assume) for Super Rock detectives, the main method of choice for figuring out the age of rocks – and subsequently, the history of the planet – is radiometric dating.

Luckily for Super Rock detectives (which I might have to start calling geologists again soon) many rocks contain what are known as radioisotopes. These are particular types of chemical elements that are unstable, meaning that they decay over time and produce radiation. The handy thing about radioisotopes is that they act basically like chemical clocks.

This clock-like property is very helpful for SRDs (Super Rock detectives) particularly because different types of radioisotopes decay at different, predictable rates. This is kind-of the equivalent to having different types of sand timers for different amounts of time, but in the case of radioisotopes, these timers can last for millions, sometimes even trillions of years.

Let's use uranium as an example: It's one of the most common radioisotopes that SRDs use. All radioisotopes have a half-life, which is how long it takes for half of the substance to decay. One type of uranium (U-238, if you want to get all scientific about it) has a half-life of about 4.5 billion years, and as it decays it ultimately turns from uranium into lead. So, after 4.5 billion years, you'll have half uranium, and half lead! Then, after another 4.5 billion years, half of the remaining uranium would have turned to lead, so you'd be left with 25% uranium and 75% lead.

Just in case any of this rock stuff is going over your head, here's the skinny: As long as a rock contains one of these radioactive 'clocks', all a geologist (sorry, SRDs) has to do to figure out the age of a rock is check to see what substances it contains. So, if a rock contains uranium and lead in equal quantities then...

(That rock is probably 4.5 billion years old.)

Strangely, for a subject that seems on the surface to be very boring (but is actually very exciting, as you now know) the subject of radiometric dating can sometimes be a little controversial.

Ooh! Can I get a...?

"Is he radiometric dating AROUND?"

Am I right, ladies?

Of course, the controversy surrounding the age of rocks is less about romance and more about... the age of rocks. I mean, say you accidentally age a rock that turns out to be a meteorite, and therefore could be older than the Earth, but mistake it for an Earth rock! That could lead you to make a major mistake in your calculations about the age of the Earth!

COULD YOU EVEN IMAGINE THE DRAMA? THINGS COULD GET VERY ROCKY INDEED!

Basically, radiometric dating (as useful as it is) has its downsides – it only works on certain types of rocks, for one thing. So what, pray tell, do our detectives do when radioisotopes are nowhere to be found?

"This rock is deep, man... it has many layers."

For all the rocks out there that don't have a handy time stamp imbedded in them, geologists have to get a bit more creative. This is when they go into serious detective mode.

As an example, let's take sedimentary rocks. (Sounds a bit like sedentary people who sit around all day watching TV and collecting dust, e.g. me.) These are rocks that form continuously over time as many different, small rocks slowly pile on top of one another. In this scenario, given the complex make-up of a sedimentary rock, radiometric dating simply doesn't cut it. Instead, the only way to age the rock accurately is by taking note of the layers of the rock surrounding it, and then comparing these layers to other layers of rock found around the world.

It's here that, I think, we reach the core issue with geology: it sounds way more boring than it actually is. It's hard not be confused by the idea of people going around the world just to... stare at rocks. Analyze rock layers. Find more rocks. Stare at those rocks. Analyze them. Compare those layers to the last set of rock layers... it's just like... man!

"Who wants to spend their life looking at boring old rocks! Eh, bro? I'd rather study the layers of that tasty cake!"

(Or whatever it is that bros say to one another?!)

However, for me this is where geology becomes fascinating not in spite of how boring it seems, but actually because of it.

I mean, all you really need to do here is look at the geologic time scale to get a glimpse at the amount of work these people have achieved. I'm not going to write it out in this book because it's incredibly dense and complicated, so just type into your search engine 'International Chronostratigraphic Chart 2015' and you'll see what I'm getting at (by which I mean, you'll be totally overwhelmed). In it, the geologists have divided the history of Earth into different units of time, which are (in descending order of duration): **eons, eras, periods, epochs and ages.**

As soon as I learnt this, I suddenly became very embarrassed that I'd been using all of these words to describe 'a really long amount of time' for most of my adult life. Like, I've been doing that for ages...

EDITOR'S NOTE — CHARLIE, THIS IS A DAD JOKE. YOU ARE NOT EVEN A FATHER. PLEASE GET IT TOGETHER.

The collective work of the geologist community around the world is a true feat of collaboration. By analyzing different rocks' layers across the planet, they've been able to map out the history of our planet to an incredible degree of precision and accuracy. It's now possible to look at a layer of rock in the side of a cliff and know exactly when that rock was formed, as well as what was happening on Earth at that exact time. This might include events such as the forming of a mountain, volcanic activity lasting millions of years, or the creation of an entirely new ocean.

It was James Hutton, the father of modern geology, who said:

"The present is the key to the past."

Geologists analyze the rocks of today to figure out what happened on this planet hundreds of thousands of years ago. But for me, the geologists are the key to unlocking the past. Maybe I'm being ignorant to how much fun the field actually is (they probably throw some *rockin'* parties!) but the feat of data analysis by these folks is truly unbelievable. I couldn't spend all my time staring at rocks, but I sure am thankful to the people who have.

So! Now that I've done my best to make fun of and then crawl back to the feet of the geologists, let's get on to what we have learnt through all their analysis...

Hell on Earth (the Early Days)

There is indeed an AWFUL lot of Earth history I could get into if I wanted to, but instead of tackling the whole shebang I just want to talk a bit about the very early days. Although actually, given how Earth's timeline is divided up, its 'early days' actually makes up almost 90% of its total history! This is known as the Precambrian eon, which spans from the formation of the Earth around 4.5 billion years ago, up until 542 million years ago, after which life finally started blooming. The Precambrian eon is divided into three, smaller eons: The Hadean, The Archean and The Proterozoic... but given that baby pictures are always the most hilarious to look at, I'm just going to look at Earth's equivalent of those:

THE HADEAN EON

Beginning around 4.6 billion years ago, the start of this eon marks the birth of our solar system. In just the same way as all the other planets, the Earth formed out of the hot, cosmic gas and dust in orbit around the Sun.

About 4.4 billion years in – once the Earth had successfully cleared out all the debris in the surrounding area of the solar system – it was finally the same size as it is today. But it wouldn't be easy to recognize this planet as our Earth, as it was essentially just a giant sphere of molten rock – there wasn't any atmosphere either, so the surface of the planet was in direct contact with space. This cold, surrounding space would cool down the surface of the Earth to form a crust, but it was never long before whatever thin crust the Earth formed would get churned up and dragged back down under the lava-like surface.

Just a quick reminder: THIS IS WHERE WE LIVE. I mean, obviously Earth isn't quite like this anymore, but still! Look down at your feet... skip back 4.4 billion years in your mind, and this is what the ground beneath us actually used to look like...

Slowly but surely, enough heat managed to leave the planet into the surrounding space, and a thin, solid crust formed to create the Earth's surface. Later, Earth formed an early type of atmosphere – although filled with extremely toxic gasses. By about 4 billion years in, the surface had finally cooled enough to make a more familiar looking place: with an iron core, oceans, and even the first land masses had probably begun to form. (So, not so much of a hilarious childhood for Earth as a fiery, terrifying one!)

THE EARTH IS MADE UP OF THREE MAIN LAYERS: THE CRUST AT THE TOP, THE CORE AT THE CENTRE, AND THE MANTLE BETWEEN THEM. THEN, ABOVE THE EARTH LIES THE ATMOSPHERE...

Core, Blimey!

Dig a deep enough hole and right at the centre of Earth you'll find the core. This is an enormous ball of metal, mostly a mix of iron and nickel, and is split up into two parts: the outer liquid core, and the inner solid core.

The deepest hole humankind have ever managed to dig reached a depth of about 12km (7.5 miles) – they had to stop because the hole got too hot for them to go any further. In order to reach the core they would have needed to travel around another 2,908km (1,807 miles) down... which begs the question: how exactly do we know anything about the core, given that we've never even seen it?

You're forgetting, of course, that geologists are **SUPER ROCK DETECTIVES** able to solve any mystery – even one as deep as this.

One thing that's easy to figure out from the Earth's surface is our planet's gravity, which allows us to estimate that it weighs around 5.9 sextillion metric tons (this is 59 followed by 20 zeros). However, the surface of the Earth isn't actually dense enough for that number to make any sense – if the Earth were the same consistency throughout, it'd weigh much less. So, in order to account for all of the missing weight, it makes sense that there must be a massive amount of material at the planet's centre – **the core!**

That in and of itself doesn't tell us how deep the core is though, so to get a more accurate understanding about what's going on down there, scientists use a pretty unorthodox measuring tool:

EARTHQUAKES

When an earthquake happens it sends a shock wave through the centre of the Earth that can be felt on the other side – and most importantly, these vibrations can be measured. What's more, they 'sound' different depending on what type of material they've passed through. So, with enough of these tremors from hundreds of powerful earthquakes across the planet, scientists have been able to use this data to figure out what the Earth looks like on the inside. However, while these earthquakes have helped us to figure out the depth of the core, as well as its consistency (liquid on the outside, solid on the inside) they're not quite as helpful when it comes to figuring out what it's actually made of. To get a good hint about that, scientists had to look back up at the stars.

Coming in at number 6 on the top ten list of the most abundant elements in the universe is, of course, iron! (Not sure why BuzzFeed hasn't covered that one yet.) Given how much iron there is out there in the cosmos, it's not actually very abundant on Earth's surface. This is the main clue that has led us to believe that, during the formation of the early Earth, most of the iron must have sunk its way down into the core. It's thought the core is 80% iron, although that figure is still up for debate. Ok, so... the Earth is full of iron. Big whoop. I mean, it's not like we can get to that metal – what exactly is it good for?

To turn the Earth into a GIANT MAGNET, that's what!

Planet Earth AKA the Magnet

When the Earth rotates, the outer, liquid core at the centre of the planet spins. This movement generates an electric current inside the planet, which in turn creates a magnetic field around it, reaching out 65,000km (40,000 miles) into space. It's this field that basically turns the Earth into a giant magnet – you'll find a field like this around any typical magnet... but, needless to say, the Earth's magnetic field is a little more powerful than those found around the magnets on your fridge.

Other than just being (in my humble opinion) **REALLY BLOOMIN' COOL**, the Earth's giant-magnet properties also create something like a force field around Earth, shielding us from the Sun's harmful solar winds. If it wasn't for this protective barrier, then our atmosphere would be damaged by the harmful radiation – the winds would basically rip all of the gases off the planet, leaving us with nothing left to breathe!

Fortunately for us, when solar winds send radiation in our direction, the pressure from the magnetic field sends the harmful particles around the Earth instead of towards it, sort of like a giant ship pushing the waves of the ocean to either side of it. Unlike my usual silly metaphors though, scientists actually use this one themselves – they refer to the point that the magnetic field meets the solar winds as the 'bow shock,' just like the bow of a ship! Also, in just the same way as you'll see a tail of waves behind a ship, the solar wind pushes the magnetic field into a tail shape, which extends 600,000km (372,000 miles) into space – much further than the magnetic field travels normally.

The magnetic field is also important for life on Earth for many other reasons too: it's how compasses work, as they align with the magnetic field and point to the poles of the planet; many animals seem to rely on the magnetic field for direction (e.g. birds, bees, and turtles) and so without it they'd be set off-course.

However, one of my favourite effects that the magnetic field produces isn't important for life at all... it's just really, really pretty. You might know this best as the Northern Lights (you can find this happening around the south pole too – it's known as the Southern Lights there) although their proper title is the Aurora Borealis.

ABOVE
THE EARTH

THE ATMOSPHERE

Usually referred to simply as 'the sky,' the atmosphere is the layer of air you'll find immediately above the rocky surface of the Earth. Unless you're reading this book somewhere under the surface of the Earth *(in which case I'm really happy to hear that 'Fun Science' has made its way down to the subterranean community, thank-you and come up soon, please)* then the atmosphere is the part of Earth that you're sitting in right now.

The atmosphere contains a mixture of different gases – it's mostly made of nitrogen (78%); it contains a good amount of oxygen, the gas we need to breathe (21%) and it also contains traces of other gases such as argon and the not-so-great carbon dioxide... but we'll get further into that a bit later.

Scientists divide the atmosphere into five different layers... which seems to be something they like to do with everything, doesn't it?

"Before I can eat this cake... let me just divide it up into its five different layers and give an explanation of each in turn."

The air in these different layers gets thinner and thinner as they go further up into space, starting with:

THE TROPOSPHERE

This is the lowest layer of the Earth's atmosphere, and the one you're sitting in right now! It starts at the surface of the Earth and goes roughly up to a height of between 8–14.5km (5–9 miles) above sea level. The height of it changes depending on numerous different factors, including the season (it's lower in winter and higher in summer), whether it's night or day, or simply just what part of the Earth it's floating above.

"It's air, man! Let it be FREE! Don't try and conform it to any strict height conformations. Just let it be."

The troposphere is easily the busiest, most happening part of the atmosphere. Almost all of the mass in the atmosphere sits in this layer – it makes up about 75%–80% of it. Nearly all of the water vapour and dust particles are found in it, which is why most clouds are found in this lower layer, and it's also the area in which almost all of the Earth's weather occurs.

This layer also tends to be the most affected by the Earth below it, given its proximity. The winds are affected by the ground – you'll find jumbled winds near 'bumpy' parts of Earth, such as forests and mountains, and smooth winds over the flat parts, like the oceans. The temperature of the troposphere is also affected. The Sun warms the Earth and this heat tends to rise up, but the heat is lost the higher up you go. It gets very cold when you get to the top, with typical temperatures of around −55°C (−131°F). This makes the troposphere unique because other parts of the atmosphere don't tend to be at all effected by the Earth below. The air also gets thinner the higher up you go, which is why mountain climbers reaching high points in the troposphere need to bring their own oxygen with them in order to breathe.

THE STRATOSPHERE

You might think the 'stratosphere' is the highest point in the sky but it's actually only the second layer up! You'll find it just above the troposphere, and it ends around 50km (31 miles) above the Earth. At the top of this layer, the air is about 1,000 times thinner than it is at sea level. This makes it a great spot for airplanes and weather balloons, so if you're flying somewhere on your summer holidays, this is the area of the sky where you will be hanging out.

It's also in the stratosphere that you'll find the ozone layer – ozone being an odd type of oxygen that is abundant in this part of the sky. The ozone actually heats up the stratosphere as it absorbs ultraviolet radiation from the Sun, which means the atmosphere actually starts to warm up again as you travel up through this layer – the opposite to how the troposphere works.

This area is nice and stable compared to the lower, rougher troposphere, which is why jet aircraft tend to fly through the lower parts of the stratosphere instead of the troposphere (which would allow them to reach their destination quicker if it weren't so windy). It's very dry too, meaning you'll find very few clouds in this layer.

THE MESOSPHERE

One layer above the stratosphere we enter the mesosphere – starting at 50km (31 miles) above sea level, and ending at about 85km (53 miles). Here the temperature starts going down, and the top of the mesosphere is the coldest part of Earth's atmosphere with a temperature of around −90°C (−130°F).

It should probably be called the 'mysterosphere', as scientists know the least about this part of the atmosphere. This is simply because it's very difficult to study as planes and weather balloons can't reach it, and satellites can't orbit low enough without falling down to Earth... so we don't have many ways to get scientific instruments into the mesosphere to take any measurements.

FUN FACTS:

· Most meteors from space burn up in this layer.

· Most of the movement in this layer is caused by waves of air that travel up from the lower atmosphere.

...and that's it. How MYSTERIOUS!

THE THERMOSPHERE

This is the fourth layer of the atmosphere, and it extends from about 85–90km (53–56 miles) up to about 600km (372 miles) above our planet, making it easily the tallest layer of all. It's also the hottest layer, with temperatures rising to well over above 1000°C (1832°F)! This is because a lot of the ultra violet radiation from the Sun gets absorbed by this layer. When it's particularly hot and full of radiation, this layer 'puffs up' which constantly affects its height.

Although the thermosphere is technically considered to be a part of the Earth's atmosphere, the air in it is so thin that it's pretty much what we'd think of as space. Most definitions say that outer space starts around 100km (62 miles), and the International Space Station and many satellites actually orbit in this area of the atmosphere. This layer is also home to another, thin layer of space (not another atmospheric layer, but still pretty important) called the ionosphere. This is a point in the sky where the Sun's energy is so strong that it actually breaks apart the gas molecules. This results in electrons and molecules floating around up there.

This might not seem particularly cool in itself, but what's neat is what scientists realized they could do with it – communicate! Back when radio waves were the cool thing, scientists could point them into the sky and they would bounce off the ionosphere – making it possible for them to communicate on Earth using radio waves.

Above this layer lies the exosphere, where the air becomes very, very thin. This is the point at which the atoms and molecules in the atmosphere escape into space – and the atmosphere ends!

Climate Change

It's weird how some people just don't want to hear about certain types of science e.g. climate change – like they're happy to know about planets and stars and the universe and all that, but throw something at them (like climate change) that has been just as rigorously tested using exactly the same methods as everything else in science... and some people are happy to call it all a lie, simply because it's information that doesn't agree with their world view.

I'm sure the majority of folks reading this book are accepting and completely onboard with the idea that climate change is a real threat, but for the minority out there who might be reading this with preconceived notions about what climate change is: sorry not sorry. All I do is follow the science. Facts are, after all, impartial.

It's also worth pointing out that I'm deliberately avoiding the term global warming, as I think it can be a little misleading. Cue all the naysayers with their:

"If it's called global warming, why is there more snow! It's not GLOBAL COOLING!"

Although, the global temperature of the planet is increasing, this can seem a little less apparent when you get into the finer details. More on those a bit later, but first let's get something clear...

THE NATURAL GREENHOUSE EFFECT

This is probably the term we hear bandied around the most when it comes to climate change, but the natural greenhouse effect isn't in itself a bad thing. In fact, without it, we'd be experiencing some of that 'global cooling' that climate change cynics seem to love to mention so much.

All of the Earth's natural, life-sustaining heat comes from the Sun. When its heat reaches the Earth, the energy passes through the atmosphere on its way to the planet's surface, with a small portion of that heat being reflected back into space. This heat energy warms up the surface of the planet, and then as the temperature of Earth increases, some of this heat is sent back into the atmosphere.

Now, if this was the final step in the process of the Sun's heat transferring to Earth, we'd be looking at an average temperature of around –18°C (–64°F). Not only would this be pretty darn unpleasant for us, but it'd actually mean that life on Earth wouldn't have been able to start in the first place. In order for that to happen, we needed a little bit of a boost.

Enter (drum roll please):

THE GREENHOUSE EFFECT!

So-called because it works in a similar way to how greenhouses keep their plants warm (although it's not quite the same). What happens is that some of the heat radiating from the Earth gets sent back into the atmosphere and then is absorbed by naturally occurring greenhouse gases, which include water vapour, carbon dioxide and methane.

Once these different gases absorb the heat energy, they start to vibrate and radiate the energy all around them – and they send around 30% of that energy back down to Earth! This increases the Earth's temperature from that –18°C (–64°F) to a comfortable 15°C (59°F), a delightful average temperature for us life forms here on Earth.

...and that's the greenhouse effect! Quite handy for us, eh? Unless, of course, you end up with more of those greenhouse gases than you can handle... And I think we all know where this is going...

THE ENHANCED GREENHOUSE EFFECT

Don't be fooled by the term 'enhanced,' as this greenhouse effect ain't better than what came before it.

Starting around the time of the Industrial Revolution, we humans got really excited and distracted by all of the new, cool, technologically advanced stuff we were inventing. So much so, that we failed to notice that our new contraptions were creating new greenhouses gases that would find their way into the atmosphere – particularly carbon dioxide. As we know, more greenhouse gases mean a higher surface temperature for Earth, and so temperatures have been rising all over the globe ever since.

Then, of course, even when we did notice that this was happening, we basically procrastinated doing anything about it until the changes we'd made were pretty much irreversible.

A BIG WHOOOOOOOOOOOOOOOOOOOOOPSIE!

Carbon dioxide, the main proponent of all of this rubbish, is currently responsible for about 60% of this enhanced greenhouse effect. It used to be that the amount of carbon dioxide in the air was perfectly balanced – it was created from natural sources such as volcanic eruptions, the decay of plants, and as a waste product from breathing animals (hey, that's us!). What was created was absorbed by planets for energy, or it was dissolved into the ocean. Unfortunately for us, even a small change in this delicate balance of carbon dioxide can have a very large impact – one that we've been inadvertently creating for the last couple of hundred years.

After burning a bunch of fossil fuels such as coal, oil and petrol, we've released the carbon dioxide that was stored in those fuels millions of years ago into the air. Although we are doing some things to reverse this effect today, much of our society still relies on these fossil fuels to run our cars and power our homes and businesses. We've also been chopping down trees in order to farm and to build, which not only reduces the amount of carbon dioxide that they could be converting back into oxygen, but it also releases the carbon stores in the trees.

As a result of all of this, we've increased the amount of carbon dioxide in the atmosphere by about 40%. We've also seen increases in other greenhouse gases such as methane which accounts for about 20% of the enhanced greenhouse effect, and has increased by about 2.5 times because of our actions.

This is the section where I'd really like to talk about all the cool ideas and plans scientists have for how we could reverse all of these effects... but because the world is the way it is and we humans simply love our fossil fuels, we still haven't come up with a long-term solution... yet.

CHARLIE'S HANDY DANDY CHEAT-SHEET FOR PROVING THAT CLIMATE CHANGE IS A REAL THING:

1. **There has been a global rise in temperature.** This might seem obvious, but it's worth hammering home. All of the scientific evidence says that global surface temperatures have risen since 1880, which ties in with the impact of the Industrial Revolution starting around 1760. There has also been a major increase since the 1970s, with the start of the century seeing the highest recorded temperatures yet! So yes, it's getting worse...

2. **The sea levels are rising.** In the last century, sea levels have risen about 17cm (6.6in). This is partly due to ice sheets and glaciers melting, increasing the amount of water in the oceans, but also this increased temperature causes something called thermal expansion, which expands the oceans themselves! Good news for everyone who lives on a hill, at least?

3. **The oceans are much more acidic.** One of the big places that carbon dioxide in the atmosphere goes is right into the sea. As CO2 levels in the oceans increase, the water becomes more acidic and the surface of the oceans has increased it's acidity by about 30% since the beginning of the Industrial Revolution, all of which is having disastrous effects on our pretty coral reefs... effectively 'bleaching' them white.

4. **We're seeing more extreme weather events.** When someone wonders why they're seeing more snow during global warming, the key thing to mention is that the increased temperature is simply evaporating more water into vapour, which has to come back down to Earth somehow! Mainly, this happens in the form of intense rain and snow storms, both of which pose very real, immediate threats.

ALL OF THIS IS TO SAY: Yes, climate change is a **REAL** problem, and one that isn't worth ignoring simply because you don't feel like separating your recycling. So maybe it's worth putting in a little effort, just on the off-chance it helps to save the planet (and us?) Or, maybe that's just me...

THE MOON IS A SATELLITE OF EARTH IN ORBIT AROUND OUR PLANET AND IS ABOUT 384,000KM (239,000 MILES) AWAY.

Its official name is simply just the Moon, which might not sound very imaginative, but bear in mind that it was named at a time when nobody knew that any other moons existed. It wasn't until 1610 when Galileo discovered four moons orbiting around Jupiter that people figured it was worth giving them proper names to distinguish them from one another. Unfortunately, our Moon never got that treatment (although you can refer to it by its Latin name, *Luna*, if you so please).

Arguably the most important fact about the Moon though is that if it didn't exist... you wouldn't be reading this book today!

Name:
THE MOON

Circumference:
10,917KM (6,783 MILES).

Mass:
0.0123 TIMES EARTH'S.

Volume:
0.02 TIMES EARTH'S.

Density:
0.6 TIMES EARTH'S.

Temperature:
123°C (253°F) DURING DAYLIGHT,
−233°C (−451°F) AT NIGHT.

Rotation:
THE MOON ROTATES ONCE ON ITS
AXIS EVERY 27 DAYS – EXACTLY
THE SAME AS ITS ORBIT.

Orbit:
IT TAKES THE MOON 27 DAYS TO
ORBIT AROUND THE EARTH.

Gravity:
0.166 TIMES EARTH'S.

Magnetic field:
LESS THAN A 100TH EARTH'S.

<<<<<<<THE<MOON<<<<<<<<<<<<<<<<<<<<<<<<<<<<<<<<<<<<<<<<<<<<<<<<<<<<<<<<<<
<<<<<<<<<<<<<<<<<<<<<<<<<<<<<<<<<<<<<<<<<<<<<<<<<<<<<<<<<<<<<<<<<<<<<<<<<<

Contrary to what you might be thinking right now, I'm actually not about to assert that the Moon was vital to the formation of life on Earth, thus enabling me to write this little science book. Although it is true that old Moony probably did help life along – its tidal forces transported heat around the early planet's oceans, making it easier for life to form – without the Moon, life on Earth would probably still exist.

Instead, I'm talking about my mother.

(EDITORS NOTE:) REALLY CHARLIE? OF ALL THE DIGRESSIONS...

My mum is, undoubtedly, one of the most supportive people (of me) on this planet. (Of course, I have no scientific data to back up this fact, and I also have an inherent bias because she is my mother... but still.) I mean, what other mum would notice that their kid is talking to themselves in their bedroom and uploading the results to the Internet, and then not only decide to encourage them to do more of it, but actively start posting YouTube videos of their own?

Basically the 'Fun Science' strand on my YouTube channel, and subsequently this book, wouldn't have existed if my mum hadn't offhandedly once said to me:

I have no idea where she got this notion from. Are people just not talking about the Moon as much as they did back in the good old days? Did the Moon's public relations team go on strike? Did we all just kind of get over the Moon once we visited it a bunch of times? (Actually, that option doesn't sound super farfetched.)

Whatever the reason behind my mum blurting her Moon concerns out, it inspired me to spread the good word about the Moon. Thus, my first ever science video: **'Fun Science: The Moon'** was born. So, let me recap some of what I mentioned in that original video, as well as giving you some tasty new nuggets of Moon information.

The Dawn of a New Moon

While we might like to think of our Moon as being 'average' among moons (probably given its incredibly generic name) our Moon is the largest in the entire solar system relative to the size of its planet. Basically, compared to how all of the other moons look next to their planets, our Moon is HUGE. The only major exception to this is Charon, which is actually the largest moon relative to its parent body Pluto... but, maybe let's not get into Plutogate again.

Is 'Plutogate' catching on yet, guys?

Usually, moons tend to form in a couple of different ways. In a similar style to how the early planets formed in a ring of gas and dust orbiting the Sun, moons can form when gas and dust in orbit around a planet clump together. Or, if you weren't lucky enough to form a moon of your own, you can always grab a nearby asteroid or two and bring those into your orbit, like Mars did (probably).

However, given the large size of our own Moon, neither of those theories really work for it, so we had to come up with something new. Something... a bit more special, we are curious creatures after all.

There are a few different theories as to how the Moon formed, but the prevailing one right now is that it was probably born out of a huge planetary collision that occurred during the early formation of the solar system. This is known as the **'giant impact'** hypothesis.

I've said it before and I'll say it again: that is an awesome name for a hypothesis.

Around 4.5 billion years ago, there was a planetary body around the size of Mars. This body, known as Theia, got a bit lost as it found itself on a collision course with early Earth. Then, as tends to happen in science when things are on a collision course, Earth and Theia collided. It's here that things got a bit messy – the resulting debris from this collision was mostly pulled into Earth, but what was left over from the smash went on to form our Moon.

It was the early solar system, guys! Try not to judge. Fun was had, mistakes were made... and moons were born.

Our View of the Moon

So, here's a major pet peeve of mine – I absolutely HATE it when, in movies and TV shows, the Earth and Moon are depicted like this:

When in fact, the actual distance between the two bodies is more like this:

For one thing, I'm sure that this error of scale has resulted in lots of us viewing this configuration on screen and thinking "Oh, I guess that's what it must really look like!" and then moving on without a second thought. I know that before I was particularly interested in science, I definitely fell into that camp, so don't worry too much if you were living under that assumption too.

What really gets me about this mistake is that, if the Earth and Moon were actually that close... **IT WOULD DESTROY ALL LIFE ON EARTH.**

The closer an object is to us, the more powerful its gravitational effect on us is. Take this book, for example: even given its small weight and size compared to say, Jupiter, the book actually has a stronger gravitational pull on you than that planet does, simply because of its proximity. So, if the Moon were in orbit super-close to Earth, then its gravitational pull would be much higher, and the effects of the tides would be much greater too – by about 100,000 times.

These more powerful tides would reap absolute havoc. There would be flooding all over the planet and tidal waves kilometres high! But not only would it affect Earth's oceans, but the force of gravity would distort the Earth, stretching and pulling on it, causing earthquakes of immense power and force. It'd also heat up the core of the Earth, resulting in terrifying volcanic activity.

Actually, having said all of that, it kind of suddenly makes sense to me why Hollywood would want to depict the Earth and the Moon as being so close together – it'd definitely make a great apocalypse movie.

The damage wouldn't end there, though, as the Earth's effects on the Moon would be much more destructive. The Earth's gravitational force would actually be strong enough that it'd tear the Moon to pieces, probably leaving quite a pretty ring system in place around the Earth, made up of pieces of the Moon.

So, if you ever see this mistake made on the big screen, just shout out...

"THAT DEPICTION IS NOT SCIENTIFICALLY ACCURATE AND WOULD ACTUALLY RESULT IN UNTOLD DESTRUCTION!"

...and I'm sure everyone will appreciate your input.

Just don't tell them that I told you to say it, okay?

CHAPTER 4

LIFE

THE ONE THING MY 'FRIENDS' ALWAYS USED TO TELL ME TO GET, INSTEAD OF SPENDING ALL MY TIME ONLINE.

Evolution

Ah... evolution. It truly is a beautiful thing. I can still remember my first time – me and my Charmander were training somewhere north of Viridian City, when my little fire lizard finally reached level 16 and evolved... into a bigger, stronger Charmeleon!

If you're anything like me (sorry if you've never played the game before – I promise we're getting to the science part), then you might have thought of evolution in terms of something improving or getting more powerful. You might even see humans as the pinnacle of said evolution. **Clearly our big brains make us the most evolved living creatures on the planet, right?**

Well, you're not too far off – but the reality of evolution is that it's not really a process of organisms getting better. Earthworms, for example, haven't been trying and failing to grow arms and legs this whole time. In fact, the earthworm is just as evolved as humans are.

Because, in reality, evolution is about one thing:

SURVIVAL.

The theory of evolution describes the process by which groups of living things – plants, bacteria, birds, bees, humans, etc., adapt and change over time in order to keep their group alive. Unlike Pokémon (I can't believe I'm still using it as an example) evolution doesn't happen in a single organism. Once a living organism is born, it's basically stuck with the genetic material that it's been given. But what about the human race and every other species for that matter? Over generations, a group's genetic material can change. In fact, it can change a LOT.

To give you an idea of how big that change can be, consider the central idea of evolution – that all of the different species living on this planet have developed from the very simple organisms that emerged on Earth more than 3.5 billion years ago. This early life was incredibly basic compared to life today – microscopic, single-celled bacteria chilling in the primordial ooze of early Earth. And these cells? You can think of them as the great-great-great-great-and-so-on-for-quite-a-long-time grandparents of the entirety of the flora (plants) and fauna (animals) that live on the planet today.

So, yes – we've changed a bit since then (just a bit).

At this point it's worth taking a moment to define what a species actually is. Now that we know all life on Earth is related, where do we draw the line? Well, by species, we just mean a group that could breed amongst themselves in their natural environment. The evolutionary theory (we have reached so far) is that all living things on Earth are part of the same family tree, and thus have much more in common genetically than you might initially think.

All of this presents us with one big question, though: if we're the descendants of these simple, single-celled organisms, then where did those organisms come from? Well, if I knew the correct answer to THAT question, then I'd probably have a Nobel Prize right now.

The best current guess is that billions of years ago, certain molecules – tiny chemical particles – emerged on Earth with a very special talent – they could copy themselves. You can think of this like a very early form of DNA, which is the stuff we rely on to reproduce ourselves (more on that sexy topic a bit later).

EDITOR'S NOTE I FIND IT WORRISOME THAT YOU REFER TO DNA AS 'SEXY'.

Now these early, special molecules may have originated on Earth, but it's also possible that they arrived from elsewhere in space and landed here on the back of a comet – scientists can't really agree on the specifics. (But just IMAGINE if it turned out to be the latter! Would that make us aliens?) However it started, over a VERY long time these tiny little things reproduced themselves and started joining forces, forming more and more complicated organisms: initially single-celled creatures, which then combined to create multicellular organisms in ever more varied and wonderful forms.

And that, in a nutshell, is evolution!

Of course none of that explains quite how it all happened... and is still happening right now. This is the process of natural selection, which was the brainchild of the one and only...

Charles Darwin (and the History of Evolution)

'MY MAIN MAN'

Just another Charlie who wrote a science book! (Although I have a feeling that *On the Origin of Species* might be slightly more renowned.)

If you were to meet a young Charles Darwin, he probably wouldn't have struck you as someone who was going to change the course of science. Growing up, he found medical lectures at university 'dull' (like most people!). Although, he did enjoy collecting beetles – and it was his passion for the natural world which came in handy when his big break arrived: he was invited on board the HMS Beagle for a 5-year journey around the world...

Fortunately for us, this trip got him thinking... Why were species so different wherever he went? On the Galapagos Islands, he noticed that the different songbird specimens that he collected varied massively depending on which island they were from. The islands were pretty similar to one another, so why did one type of finch have a thick, large beak, while another type had a long, thin one?

The answer, he discovered, had to do with the differences in their environments. The big-beaked finches had a diet consisting mostly of nuts and seeds, so their large beaks were super useful to crack nuts apart to eat them. The long-beaked finches however had a taste for grubs, so they were able to poke their thin beaks into the ground to extract them. In both cases, their physicality had changed to best suit their environment. But this wasn't by design... this was just nature. So what was causing it?

Now, by this point in history, scientists had come to the consensus that the creation of all of the various species on Earth seemed to have occurred at different points in time, not in one sudden, miraculous event... But this was as far as they'd got.

It was then that Darwin figured out the last piece of the evolutionary puzzle – a realization that another naturalist, Alfred Russel Wallace, actually came to at around the same time: the process of... **natural selection**.

As you might already know, you are made out of a mix of genetic material from your biological mother and father: 50% from him, and 50% from her. At least, that's how it's supposed to work anyway, but the process of making a brand new living thing doesn't always go perfectly. Everyone makes mistakes! Even Mother Nature.

From time to time, during the process of copying genetic material, little errors get made and new, unexpected traits appear. Sometimes these traits are actively detrimental but a lot of these mistakes don't really make any difference. You might have ended up with a double-jointed elbow, which, yes, is a fun thing to show people at parties, but isn't necessarily going to help you survive an Ice Age.

But here's the rub: every now and then, Mother Nature makes a little copying error that actually results in a new, awesome trait appearing! Thicker skin, for example, which would definitely be useful in an Ice Age. These bonus traits could be anything – bigger teeth, longer legs or really pretty feathers (seriously). So long as it gives you a survival advantage in your specific environment, it's worth having.

It's this process that truly is the key to how evolution functions. If you've got thick skin, you'll do better in a cold climate, which would make you more likely to breed and then pass that new trait onto your offspring. It's an incredibly slow process (like I said earlier, it's taken us 3 BILLION YEARS to get to where we are today) but as these new traits are passed down the genetic line they become the new attributes for an entire species, while the unhelpful attributes (like a mouse who had a habit of shouting **'EAT ME!'** for no apparent reason) are stamped out.

Given enough time, these new traits build upon one another to create organisms that look and behave completely differently to their ancestors. This is still a random process, though, which is why even animals adapting in very similar environments can end up looking completely different – like the apes which evolved on the mainland of Africa **(that's us!)** VS the lemurs that evolved in Madagascar. We both share a common ancestor, but the African island split off from the mainland 135 million years ago, and thus it resulted in two entirely different strains of life growing in tandem with one another. It's a beautiful process, especially when you consider the sheer wealth of life we find on this planet today.

It's pretty amazing to think that all life on Earth – every species on the planet – is ultimately related in one enormous family tree, sharing common ancestors all the way back in time. While some people might like to see the human race as 'special' compared to other life, I find the fact that we are all so similar quite humbling. We may all look different, but we really do have a lot in common – **for example, did you know we share about 50% of our DNA with bananas?**

As Darwin said: *"From so simple a beginning endless forms most beautiful and most wonderful have been, and are being, evolved."**

Or as I like to say: *"Please don't eat that banana – he's a distant relative of mine."*

FUN FACT: Even though evolution has a habit of changing organisms, some creatures have adapted so little over the vast expanses of time that they're referred to as 'living fossils'. The coelacanth fish was thought to have become extinct some 80 million years ago, but in 1938 a living one was discovered in the Indian Ocean which looked incredibly similar to its fossilized ancestors. It turns out that sometimes a species doesn't need to adapt – it was surviving just fine in its original form, so it stayed that way.

*Darwin, Charles, *Origin of Species* (Wordsworth Editions, 1994)

charlie loves

The Dinosaurs

One of the earliest forms of life on Earth were the trilobites – a funny-looking marine dweller, sort of like a crab with a tail, which were perfectly adapted to swim in the ocean about 500 million years ago. After them came the sea scorpions, then early fish, giant insects and early reptiles… Eventually leading to their slightly more **famous** and **attention-grabbing** descendant: **THE DINOSAURS.**

So, why did the dinos get all the glory? The earliest known discovery of these land-dwelling reptiles happened some 3,500 years ago in China, when people found what they genuinely thought were dragon's teeth! (Not a bad guess, really.) Later on, discoveries of dinosaur bones were mistaken for the skeletons of giants. (Not quite as good a guess…)

In the early 19th century a British scientist, Gideon Mantell (not to be confused with my cat, Gideon) correctly identified some fossilized teeth as having belonged to a gigantic, plant-eating reptile. Not long after that, in 1842, the British paleontologist Richard Owen came up with the word 'dinosaur', which comes from the Greek *deinos*, meaning 'fearfully great' (as in awe-inspiring) and *sauros,* meaning 'lizard'. And thus, the GREAT-LIZARDS were discovered!

The dinosaurs lived between 230 and 65 million years ago, in what is known as the Mesozoic era. Scientists split this up into three different periods: the Triassic, the Jurassic and Cretaceous – but really, the main thing to remember here is that these beasties really did rule the Earth for about 160 million years. To put that in context, the human race has only been around for about 200,000 years, so compared to the dinos we've hardly even made a mark yet.

What was their secret for survival? They had **GREAT** posture, of course! Well, almost. In those prehistoric times, standing up straight was what gave the dinos a huge advantage. Unlike the squashed-looking, sideways-sprawling legs that today's lizards and crocodiles have, the dinosaurs' legs were straight and perpendicular to their bodies, meaning they could move much faster. This gave them an advantage: whether they were running away from predators or running after things they wanted to eat, they were able to survive much more effectively.

Currently, scientists think there may have been more than 700 different species of dinosaur. The earliest seem to have been quite small – one was the size of a cat, another the size of a pony... (now I'm just thinking about riding dinosaurs or keeping them as pets.) However, at the end of the Triassic period, some mysterious mass extinction killed off many of the other big land animals, allowing the little dinos that survived to take over the planet.

From here they got bigger and bigger, eating the bountiful vegetation that covered the forested, rainy land. Meanwhile, Earth's giant, single land mass known as Pangea was gradually moving apart – isolating groups of dinosaurs which became more varied as they were left to their own devices. This accounts for why fossils of maybe the most famous dino, the Tyrannosaurus Rex, have only ever been found in North America, with no evidence of it ever spending time on what would later become the British Isles.

Now, despite what you might have learnt from *The Flintstones*, dinosaurs and humans never actually walked the Earth at the same time – in fact, there were a good 65 million years between us and them. Still, if we were to travel back in time and hang out with them, it might not be quite as intimidating as movies have led us to imagine. The reality is that most of the dinosaurs, although certainly not all of them, were probably about our size or smaller. The only reason we know about so many more of the big ones is that their larger bones were better preserved as fossils over the millions of years that they lived.

However, the average size of the dinos isn't the only misconception about them that still persists today. Although dinosaurs were a type of reptile, they might not have been cold-blooded like modern reptiles. It's likely that some carnivorous dinos, and even some herbivores, were warm-blooded, meaning that they could maintain a steady body temperature independent of their surroundings. Also, most dinosaurs were probably plant-eating, although if you're imagining these vegetarians as being any less formidable than their carnivorous counterparts, then definitely cast those notions aside. *Diplodocus*, for example, a long-necked veggie-eater, was about 35m (115ft) long, weighed between 10 and 16 metric tons, and is thought to have been able to whip its tail so fast it could break the sound barrier and boom like a canon! And don't forget that its tail spins could break bones and slash through flesh... **(Clearly, vegetarians DO kick ass.)**

Unfortunately for the dinosaurs, though, while they did definitely have a decent innings, their reign eventually came to a violent end. In a similar fashion to how the last mass extinction event allowed the dinos to take over in the first place, about 65 million years ago, at the end of the Cretaceous period something brought about their ultimate demise. What seems likely is that some huge natural catastrophe changed conditions so drastically that dinosaurs failed to adapt and survive.

So Whodunnit?

Whatever killed the dinosaurs took place over a relatively short period, and it caused many other creatures to die out too. A lot of scientists blame a giant asteroid which slammed into the Earth about 65 million years ago, triggering an enormous disruption to conditions around the planet. They've even found the 177km-wide (110-mile-wide) crater it created off the coast of Mexico. Other scientists point the finger at some devastating volcanic eruptions in India that happened around the same time, so it was probably the combination of these two events that ultimately wiped them out. In either case, it probably wouldn't have been a good time to be alive: just imagine terrible tsunami waves roaring around the world, while the asteroid impact and volcanic eruptions threw up enough dust and debris into the sky to darken it for years, depriving plants of the sunlight they needed to live, thus starving the dinos of the food they needed to survive.

This mess was handy for us humans though – or at least it was for our predecessors. The new lack of powerful rivals for food and territory meant that there were gaps in the ecosystem (the natural world) and so birds and mammals, including our ancient ancestors, were able to fill those spots, survive and thrive. The dinosaurs, on the other hand, vanished from the face of the planet...

(Well... not exactly.)

While humans have definitely taken over the role as 'rulers of the Earth', the dinosaurs have not entirely gone – they still exist in the form of their descendants. If you'd like to meet the T-Rex's closest relative, just look up! The majestic pigeon you see on the roof eating a tourist's lunch – along with every other bird from hummingbird to ostrich – is now thought to be the closest living relative to the dinosaurs.

Basically, turkey dinosaurs contain a little more real 'dino' than you would expect.

FUN FACT: It's thought that a male T-Rex's reproductive organ could have been up to 3.6m (12ft) in length. Despite those tiny hands!

Reproduction

That's right, kids – it's time to talk about getting busy! But before we get into the birds and bees... let's think about: **reproduction.**

Why do we even need it? The thing is, until we can all turn ourselves into immortal robot cyborgs, or upload our consciousness into a computer, ultimately our days on this Earth are numbered – and that's the case for every other living thing too.

WE ARE ALL GOING TO DIE SOME DAY.

EDITOR'S NOTE) DON'T FORGET THIS IS SUPPOSED TO BE A 'FUN' SCIENCE BOOK.

Ultimately, it's up to us to replace our numbers with some new folks so that the whole 'life on Earth' thing doesn't come to a screaming halt. That's the logistics of it anyway, but luckily for us we don't find the act of reproduction to be too much of a chore – and that's no coincidence. Given that reproduction is absolutely essential for life to continue, it helps if living creatures feel an overwhelming urge to reproduce.

Us, humans, happily go in for sexual reproduction, which is where two parents pool their genetic material to produce a baby that combines elements of both of them. To make sure that the child doesn't end up with double the biological material that they need, the parents combine their gametes, which are the special sex cells containing half of the usual amount of genes a cell normally contains. You might have heard of these 'gametes' already: the male gamete is the sperm, and the female gamete is the egg. When these two cells join together that's what we call fertilization, and then you've got a fertilized egg, ready to grow into a unique living organism! It'll resemble both parents, but won't be identical to either of them (and that's sex, baby).

HOWEVER, SEXUAL REPRODUCTION ISN'T THE ONLY WAY TO CREATE OFFSPRING – SOME ORGANISMS CAN DO IT ALL BY THEMSELVES! ASEXUAL REPRODUCTION ONLY INVOLVES ONE PARENT, AND YOU CAN THINK OF IT KIND OF LIKE AN ORGANISM CREATING A CLONE OR A 'MINI-ME' OF ITSELF WHICH IS GENETICALLY IDENTICAL TO THE PARENT.

The use? Well it's possible to do this anytime, anywhere, without any need to spend lots of time 'in da club' searching for a mate. So it's worth asking... why do we need sexual reproduction, given that there's an alternative option?

Not to sound like I'm going on a misandric rant or anything but... why do we need men at all? What's my willy good for? Well, the answer is simply in what that combination of two different sets of genes can do for the survival of a species.

The Genetics of the Birds and the Bees

So, you've got a sperm and an egg fusing with each other to create a fertilized egg – the beginning of a brand-new being. But let's look a little closer, at the genetics of it all. How exactly do you inherit particular traits from your parents?

Each sperm and egg contain unique genetic material: they are like two halves of a biological blueprint. The blueprint covers everything from how to make cells, tissue and organs, even how to grow hair that will form into that annoying cowlick that you have to wet down with water every day. It's all planned out in advance.

The chemical that contains all of these instructions is known as deoxyribonucleic acid... (which is why everyone just calls it DNA). One single gene, which is the basic unit of genetics, is simply a section of that DNA that contains the code to make something specific in the body. Basic characteristics (like the colour of your eyes) are controlled by just one gene, while other, more complicated instructions (like how to even make an eye in the first place) are contained in many different genes.

Everyone gets two copies of each gene – one from each parent – with different forms of the same gene being called 'alleles'. These two traits won't be identical, and the test of which of these features appears in the organism is based on whether or not the allele is dominant or recessive. A dominant trait will appear no matter what **(it's the more outgoing, less introverted gene)** while a recessive trait will only appear if the matching gene from the other parent is recessive too **(it's like the shy, silent type that will only go out to a party if it knows its friends are going).**

The allele for blue eyes, for example, is recessive, while the one for brown eyes is dominant – so you'll only show off some big blues to the world if you get the allele for blue eyes from both parents. If you get one for brown and one for blue, you'll have brown eyes. And two brown? Well, that'll be brown again.

The amazing thing about this is that almost every one of your body's cells carries the ENTIRE set of DNA instructions needed to make your entire body.

This is known as your genome, and everyone's genome is completely unique to them – which is why it's so handy in police investigations: if you leave even the tiniest speck of skin at a crime scene, it's as though you accidentally dropped your birth certificate for the police to find.

All of this DNA data comes neatly packaged in long, tightly twisted structures called chromosomes. Humans have 46 of these, arranged as 23 pairs in each normal cell. The neat thing about this is that when a sperm and egg combine, this process matches 23 unpaired chromosomes from your dad with 23 from your mum. The chromosomes you get from each of their pairs is a completely random process, which means the genetic traits you inherit is random too.

Your biological sex is just one of the many characteristics settled by this luck of the draw – just one of the 23 pairs defines whether you end up male or female. So, if you get two X-shaped chromosomes from your mum and dad then congrats, you're a baby girl – and if you end up with one X from your mum and one Y from your dad, then you'll end up as a little bloke.

FUN FACT: When you're growing in your mother's womb, your body won't know right away if you're male or female, and so it'll start developing nipples just in case. And that, everyone, is why blokes end up with useless nips.

The gene-picking process is even more random than that, though – during meiosis, which is a special type of cell division that produces the egg and sperm, each parent's DNA is shuffled to add even more variation. Then, add to that any mutations that occur naturally in genes, and you're bound to end up with someone completely unique at the end. This is why brothers and sisters end up different every time, even though they have exactly the same parents.

The real crux here is why this shuffling of genes happens in the first place – what is the reason behind every individual being so... individual? Well, it all comes back to the wonder of evolution. If we were to produce a genetic clone using asexual reproduction from just a mother, the chances of that offspring surviving is exactly as likely as the mother's chances.

When you combine different sets of genes together and throw a bit of random mutation into the mix, you're giving your offspring the chance to be better than yourself, which means they have a greater chance of survival!

And that, my friends, is why we still need blokes. For now, at least.

The Many Wonderful Ways of Mating (AKA Doing 'It')

So, that's how reproduction works on a genetic level, but what we haven't gotten into yet is the jaw-dropping variety of ways life goes about the process. As humans, we like to think that our method of mating must be the normal way, but look a little closer into the wonderful world of reproduction and you'll see that 'normal sex' is a very subjective term indeed.

Firstly, as human beings we don't achieve sexual maturity – the ability to reproduce – for many years after birth. (Heck, even once we do reach it, we tend to want to wait until we've settled with someone before we get down to business.) Then, when we do create offspring, we don't usually make that many of them. Humans create only a few sprogs each as this way we can properly look after them which will (hopefully) ensure they survive to adulthood and beyond.

FUN FACT: For a Galapagos tortoise, it's completely normal to be a 40-year-old virgin, as that's the age they reach their full adult size.

For other creatures however, the offspring turnaround can be much higher – they practically churn out babies… but then they tend to be much less successful at keeping them alive past their first year, month or even day. The fruit fly, for example, is ready to do the deed after about 2 weeks of being alive, and can then produce up to about 900 young a year. However, most of them will be eaten or otherwise meet a sticky end – which is good for us as, if it wasn't for this, we'd be overrun with them.

Then there's the tricky question of monogamy – whether to mate with one partner exclusively or not. This is a headache even for humans, but it's definitely something that pops up in other parts of nature too. It's most often seen in creatures that share parenting duties, be that feeding or protecting their shared young. Pigeons like to stick with their partner for a lifetime, while other birds, like emperor penguins, might just stick with one guy or girl for the mating season – a bit like a summer fling.

Although monogamy for humans means being totally faithful to one partner, in nature it's not that simple. Lots of animals with long-term partners hedge their bets (in genetic terms) by having other mates just in case they might produce more successful offspring. So, while about 90% of birds are socially monogamous – sharing their territory, acting like a happy couple, visiting the in-laws (maybe not that last one) on average 30% or more of the baby birds in any nest aren't the offspring of the male bird who is helping to look after them, but are actually from some rival bird. So, at least as far as birds are concerned, stepdads are pretty common!

Still, birds are much more likely than mammals to be socially monogamous – only 3% of mammals are monogamous, for example, European beavers, who are devoted joint parents and stay together until one of them dies. Whereas, other adult animals all help to raise each others' offspring, which is what meerkats do – their alpha female, who has established herself as the top lady by fighting her way to dominance, is the only one who gets to breed, but everyone helps look after her young.

(I'm sure that if this was how humans operated, we'd all be taking care of Beyoncé's children right now.)

Often, the set-up is that the alpha male mates with multiple females (who just get stuck with him), as is the case with lions, hippos and monkeys. But there are – although it's rarer – communities where the female mates with multiple males, like the honeybee. There's also always the option of a mad orgy free-for-all – herrings, for instance, collect in huge mating shoals which produce so much sperm the water looks like milk, fertilizing millions of eggs at once!

Then there are the really odd arrangements. For a long time, scientists were very confused that they never found any male angler fish, which is a disturbing-looking creature living in the depths of the ocean. All they ever came across were female angler fish, which often seemed to have a parasite attached to them... until eventually they realised that the 'parasite' was in fact the tiny male angler fish that had bitten and then bonded to the female – a dead sack of semen, ready for whenever she wants to reproduce.

After all this mating, fertilized eggs might be left to drift in the sea, nursed in a nest, or be kept safely within the body. (See! Everyone's got different ideas about child-rearing.) Then, depending on whether the parents are going to stick around to look after them, animals are born at different stages of development – some are like tiny adults, ready to take care of themselves; others, like human babies, are very reliant on their older relatives (although how hilarious would it be to have a tiny yet apparently fully-competent man pop out at childbirth).

Plants offer a whole separate world of sexual excitement. They can reproduce either sexually or asexually in a variety of different ways... **BUT THE VARIATION DOESN'T EVEN END THERE!** (And seriously, I'm only scratching the surface here.) In just the same way that a flower can have both male and female parts, making them hermaphrodites, some animals have both sets of genitals too. These include worms, snails, slugs and barnacles. Notice anything they all have in common? Well, they're either incredibly slow-moving... or they never even move at all. This is why being sexually flexible is such a plus – it means that when they do finally come across another member of their species, they can always make the most of that relatively rare opportunity to reproduce.

My favourite example of this is probably the leopard slug, which boast HUGE blue penises that grow out from behind their heads. A mating pair will hang upside down from a string made of their own slime attached to a tree, and then will wrap around one another to fertilize each other's eggs – forming an image that looks oddly like a blue flower. They'll stay like this for about an hour, and then drop from the tree and go their separate ways.

And then there's the clownfish which makes things even more confusing as they will actually switch sex from male to female, depending on whether numbers are right for breeding.

Not to ruin your childhoods or anything, but if that sad scene at the start of *'Finding Nemo'* were to happen in real life... Nemo's dad Marlin would change into a female, and then mate with Nemo. That's nature, kids!

The final option for animals needing to reproduce is the virgin birth – which *is* a real thing! At least, among some animals like the Komodo dragon, a powerful lizard that can grow up to 3m (10ft) long. They're not hermaphrodites, but they can reproduce on their own by fertilizing their eggs without sperm in a process called parthenogenesis. In one well-documented case, a Komodo dragon living at Chester Zoo, England, called Flora, had never been kept with a male yet still somehow managed to lay a clutch of eggs that hatched. After a bit of DNA testing, it was found that each egg was fertilized by another just-forming egg cell, providing a complete and unique genome for the baby Komodos. So, why reproduce this way, given the need for genetic diversity? Well, it's at least a handy technique in an emergency. The Komodo dragon lives on Indonesian islands, and with this unusual ability, if a tropical storm ever washed it up on an island alone, it could start a whole, new community all by itself.

As for simpler organisms, such as bacteria or even the relatively more complex sea anemones, they tend to reproduce by a process called fission: after a growth spurt, one bacteria splits into two organisms that are genetically identical. And then there's fragmentation, as seen in starfish, where a body breaks into two parts. This can either happen intentionally, or even if something bites the starfish in half, each part should still be able to regenerate into a healthy animal.

So, if you ever find yourself worrying about whether or not your sexual desires are normal, take solace in the fact that there's really no such thing – in the name of creating offspring, nature will try just about anything.

Bioengineering

WITH HUMANS BEING HUMANS, NOW THAT WE KNOW THE INS AND OUTS

EDITOR'S NOTE *I REALLY HOPE THAT WASN'T A DELIBERATE PUN, CHARLIE.*

OF HOW REPRODUCTION WORKS, WE'VE NATURALLY STARTED TO MESS AROUND WITH IT OURSELVES. BIOENGINEERING, JUST LIKE IT SOUNDS, DESCRIBES HOW SCIENTISTS ARE TRYING TO ENGINEER THE BIOLOGICAL WORLD IN NEW, DIFFERENT WAYS. MORE SPECIFICALLY — AND HERE'S THE PART WHERE THINGS CAN START TO GET A LITTLE CREEPY — THERE IS ALSO GENETIC ENGINEERING, WHERE SCIENTISTS CREATE NEVER-SEEN-BEFORE VERSIONS OF LIVING THINGS.

The secret to this lies, once again, in DNA – all it takes is a little fiddling about with some of the codes of life, and you can start to create something new. Scientists have already managed to genetically engineer goats who produce milk containing a drug that helps people with blood-clotting disorders. Bacteria and yeast have also been engineered so that they produce a substance very similar to human insulin (a hormone that regulates blood sugar) to be used by patients with diabetes. With this in mind, we could one day be able to vaccinate ourselves against all diseases simply by eating a genetically modified super-banana! (Which is also a superhero that happens to be a banana, obviously. Kind of like this...)

Clearly, it's an area of science that is capable of real-life miracles... But how far is too far? In Japan, for example, scientists have been breeding genetically modified mice that are prone to passing on mutations, just to see what might emerge. Researchers have already created pigs that glow in the dark – by adding a gene for bioluminescence – natural light – from a jellyfish.

Perhaps it's no wonder that people get worried about science 'playing God' – life is incredibly complex, after all. However, as is the case with any pursuit, it's not down to science to decide what's right and wrong, but instead that quandary lies in the hands of the people who are playing around with these new techniques. Let's hope, then, that we'll see more super-bananas and less glowing pigs (yes, this is still a science book).

Cloning

This is, truly, the part of science where real life starts to meet science fiction. Cloning definitely sounds quite sinister, but on its own it isn't really anything to be worried about. For example, an identical twin is just a natural clone, produced when a fertilized egg splits in two, creating two organisms with an almost identical genetic make-up. Scientists have cloned everything from genes, cells and tissues to entire animals, including a sheep!

Maybe you remember hearing about her? The famous Dolly – the first mammal to be cloned from an adult cell – began her life in a test tube. Dolly's entire genome was taken from a mammary gland cell of a 6-year-old sheep, which was implanted into the egg cell from another sheep. Then came the 'Frankenstein' moment (quite literally) as the egg was given electric shocks to trigger the process of cell division, and so baby Dolly began to develop. In the decades since, scientists have cloned everything from cats to monkeys in much the same way. So what's the use? Well, one great way to utilize animal cloning might be to make copies of endangered species, to help build up their numbers... but it's hard not to think of this area of science as fitting in the 'let's just do it because we can!' zone.

FUN FACT: Given that Dolly's entire genome came from a sheep's mammary glands (sheep boobs), she was named accordingly. As the fellow who led the team of Scottish scientists behind her birth said: *"We couldn't think of a more impressive pair of glands than Dolly Parton's."*

Fortunately, no one seems to have gone so far as to clone a human being... yet. Scientists in South Korea in 1998 claimed to have done so with a human embryo, but then stopped the experiment when it was just a group of four cells. It's actually much trickier to clone humans and other primates but there is enormous interest in using cloning technology to obtain human stem cells – cells that can develop in pretty much any body tissue – which could be used to treat diseases. This process would involve harvesting the stem cells from a cloned human embryo, which would then be destroyed... so it's definitely a controversial topic.

However, if I was to clone my own cat Gideon, the sorry truth is that I'd probably end up with an animal that didn't look much like him at all. CC, the first cat to be cloned, had a totally different coat to her clone 'parent' Rainbow; CC's tabby-and-white coat had none of Rainbow's orange fur. As it turns out, the coat colour genes in cats are randomly activated in some skin cells and not in others, so it's incredibly unlikely you'll end up with an animal that looks the same at all. In fact, the differences between CC and Rainbow pointed to other ways genetics don't dictate all our characteristics – CC, having been handled a lot more by people at an early age, was more curious and outgoing than the more cautious Rainbow.

It's nice to know, then, that – although Gideon won't be around forever – he would at least remain completely unique, even if I was to clone him!

CHAPTER 5

THE

BODY

I know I told you this earlier,
but I reckon it needs repeating:

YOU ARE A STAR!

...Or at least, the atoms in your body (atoms being the teeny tiny particles that make up ALL matter) were formed in the hearts of stars billions of years ago – which kind of makes *you* billions of years old too!

However, to look at it another way, you're also pretty simple. Basic, even.

Try not to be offended, I mean this purely in atomic terms. About 63% of the atoms that make you up are hydrogen, which is the most common element in the universe. This is because most of the human body is made up of water (H_2O) molecules. The oxygen atoms are bigger, though, so if you look at your composition in terms of mass – which is the amount of matter, or stuff, that a thing contains – you're actually 65% oxygen.

Hydrogen and oxygen aside, there are four other chemical elements that account for the other 99% of your mass: carbon, nitrogen, calcium and phosphorus. You'll find traces of aluminium and silicon inside you, too (or possibly more than just traces of the latter if you went down the plastic surgery route). But even given that these different elements are relatively simple in their basic states, these tiny building blocks are all you need to create a system of huge complexity that is your body.

FUN FACT: Recently, a researcher decided to work out exactly how much a human body would be worth if you were able to extract and sell all of its various elements. His conclusion? Well, at the time of writing this book, it's about £110 ($147). Talk about 'life is cheap!'

Although the atoms that constitute you are incredibly ancient, don't expect them to stick around. Because your body is a living thing (see, don't say I don't teach you anything new) it's constantly repairing and rebuilding itself. For instance, your entire skeleton is completely replaced every 10 years, and some cells are renewed as regularly as every 2 weeks.

It's a similar story closer up too – studies have revealed that the turnover of atoms in our bodies is rapid; we absorb new ones from our food, drink and even the air that we breathe. In just a week or two, all the hydrogen in your body will have been replaced by other hydrogen atoms, and if you wait just over a year, almost all – about 98% – of the atoms that make you up will have been replaced by other atoms.

While in one way you're billions of years old, to look at it another way, your body is practically brand new! Really, you can think of our bodies as a living pattern. So, while the stuff we're made of changes, the way in which it's all arranged stays pretty much the same.

I mean, can you even consider yourself 'real' if you're nothing but an ever-shifting conglomerate of matter... a living blueprint? (I've basically just accepted at this point that the 'fun' part of *Fun Science* actually means 'terrifying'.)

Cue everyone reading this suddenly having an existential crisis...

Regrettably, I'm not really equipped to help you to find your 'true' self (I'll save that for *Fun Philosophy*). Instead, how about I distract you from all this identity terror by showing you some blood and guts?

The Bodily Systems

So… let's zoom out from the atomic perspective (phew!) and look at ourselves not in terms of what we're made of, but how we're put together. Your cells, the building blocks of life, are arranged together as tissues – say, muscle or brain tissue – which is simply just lots of similar cells working together each with a particular job. Different types of tissues make up the body's various organs, such as the heart, brain or stomach, which again each serve a specific function. All of these organs work together in what are known as bodily systems, each with its own specific role necessary to keeping the body running.

There are a few ways to divide up the body's different systems – none of them work completely in isolation after all, as they're all part of the whole – but I'm going to separate them up like so:

1. THE SKELETAL SYSTEM
AKA your 'real' inner-strength

As human beings, we hide our skeletons away in our bodies, which is called an endoskeleton (*endo* meaning 'within'). The name skeleton comes from the Greek *skeletos*, which means 'dried up', and it's easy to think of your skeleton as some fixed, dead object: sort of like an internal coat hanger for your organs.

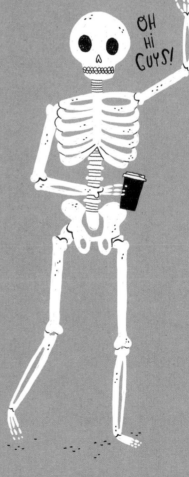

It's true that, without the skeleton's much needed support for our muscles and other tissues, we'd be shapeless, fleshy blobs. As well as helping to protect our soft, vulnerable insides and allowing us to move about – crucially, our skeletons are very much living things. We make new bone all the time, to repair and strengthen the existing bone tissue.

In total, there are 206 bones in an adult skeleton and more than half of them are found in your hands and feet, forming the incredibly intricate structures that allow us to make an enormous variety of movements. So, next time you do a thumbs up, think about the fact that it's taking a whopping **27** different bones to make that happen.

Most of our bones stop growing once we've reached adulthood, but weirdly, scientists have discovered that the human skull continues to grow as we get older: the forehead moves forward while the cheekbones move back, which contributes to the sagging seen when we age. That may mean that, in the future, those trying to stop the march of time will opt not just for a facelift on their soft tissues, but a whole bone overhaul... (Rather you than me.)

FUN FACT: We all know that, as we get older, we lose our baby teeth and our adult teeth grow through... but did you know that all children are born with their adult teeth already in place? If you were to take a child's skull and chip away above where the teeth are, you'd discover the adult teeth just sitting there, waiting to grow through!

EDITOR'S NOTE CHARLIE THIS IS WAY TOO DARK, WHAT ARE YOU THINKING?

Every bone in your body is connected to another one somehow, except one: the hyoid, a bone in the throat which works with your voice box and tongue to help you form speech sounds. A joint is simply where two bones meet, held together by ligaments – there'd be no point in having all your bones fused together, as that way you wouldn't be able to move!

Which brings me neatly onto the skeleton's best buddy...

2. THE MUSCULAR SYSTEM
AKA let's get a movin'

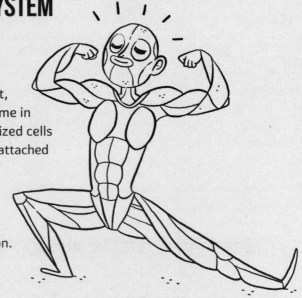

Obviously a skeleton is pretty pointless unless you've got a way to shake your bones about, which is where the muscles come in – these are made up of specialized cells called muscle fibres. They are attached to the bones by tendons, and when the muscle fibres contract (squeeze together) they pull on the bone, and so the bone moves in that direction. Then of course the question becomes, how do you get the bone back to where it was?

Well, it's not as simple as just expanding the muscle that contracted before – instead, muscles need to work in pairs.

Try this with your forearm now – extend your arm, and then flex it like you're pretending to be one of those body builders on Muscle Beach. Not only will you probably look super cool ('Look at them! They're reading a science book AND they're working out') but what's happening is that when you're moving your forearm up, you're contracting your biceps. Then, when you extend your arm back out, you're relaxing your biceps, and its opposing number – the triceps at the back of your arm – is contracting instead. Each set of muscles has a happy partner just like this – equally sharing the load of the bone that needs moving.

FUN FACT: The word muscle comes from the Latin meaning 'little mouse' presumably because back then they had little mice to move their bodies about for them.

It's not just your skeleton that gives your muscles something to move, though. Slow, involuntary muscle contractions in the walls of intestines propel food through the digestive tract (and, eventually, push the stuff you don't want out the other end).

Then of course there's all the expressive facial muscles, such as the circular 'kissing muscle' that lets you close your mouth and purse your lips. (These days, that muscle is probably used mostly for pulling the perfect selfie pout more than anything else.) Given all of these bits and pieces, it's no wonder that the muscles in our bodies account for around 40% of our weight!

Petition to rename the kissing muscle the 'selfie muscle'. Anyone...?

3. THE CARDIOVASCULAR SYSTEM
AKA the work-out part of your body

If you've ever done cardio down at the gym, this is the system you've been working out. It's also known as the circulatory system, and it's how your heart – which is the body's most powerful pump – circulates blood around your body.

Here's a quick test – put your hand on your heart. (*And take a pledge to the Church of Charlie...*)

No, but seriously, just take a moment to feel for where your heartbeat is... Ok, think you've found it? Well, most people when asked to find their heart would point slightly to the left of their chest, but (hand on my heart) it's actually right in the middle – it just feels like it's on the left because the largest part of it lies there.

The heart does the most physical work of any muscle in the body, but it's important to treat it carefully – both physically and romantically, because heartbreak is a REAL THING. Scientists have found that 'stress cardiomyopathy', a condition where the heart muscle weakens temporarily, can be brought on suddenly by an emotionally stressful life event such as divorce or even a tricky break-up.

The adult heart pumps blood through a network of different sized blood vessels – if they were all laid out end to end, they'd stretch for more than 160,935km (100,000 miles) – enough to circle the planet twice! It's strictly a one-way system, too – it contains different sized tubes called arteries, arterioles and the smallest, hair-thin capillaries – so tiny that sometimes they can only fit one blood cell at a time. These carry blood away from your heart to organs, tissues and cells, and eventually that blood is delivered back to your heart by veins and smaller venules. This set-up means that blood can reach your organs and tissues, delivering oxygen and other nutrients to every cell, and whisk away carbon dioxide and waste products. By the time the blood reaches your veins, though, the speed of that blood has slowed down considerably – which is why those tubes tend to sit closer to the surface of the skin (it's safer, just in case they open!).

While we think of blood as being red, that colour actually just comes from the red blood cells that travel in it. In actual fact, blood is mostly made up of a yellowish liquid called plasma, which is water that contains some salts, sugar and a few other things. Floating about in this plasma you've got different types of blood cells – the main and most common being red blood cells, which are dinky little disks that carry oxygen around the body. These are actually made in bone marrow, which is the soft tissue inside your bigger bones. You can think of these fellas like little truckers, delivering the oxygen that keeps the body going – and there are loads of them inside of you too. You've got far fewer white blood cells, but they're just as crucial – you can think of these like the blood's police officers, they're there to fight off infections. And then there are platelets – almost like blood doctors – who help the blood to clot when you're injured, sealing off wounds and reducing blood loss.

4. THE INTEGUMENTARY SYSTEM
AKA we are not just skin deep

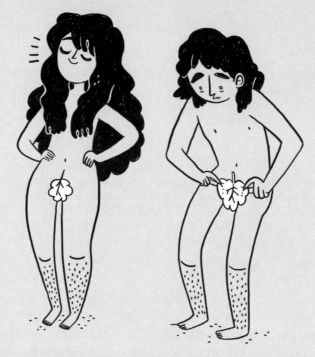

Your 'integument' is your natural covering – although most people would just refer to this as the skin. It might not seem like it, but this whole system is a bodily organ in its own right. It's made up of your fleshy shell as well as your hair, nails and exocrine glands, which produce sweat, wax and oil.

FUN FACT: Each hair on your head grows about around 1cm (⅓in) a month, meaning you produce about 16km (10 miles) of hair a year! It's also affected by the seasons – we shed our hair the most in late summer and autumn.

Your skin is actually your largest organ, weighing in at around 4kg (8lb) in total (although it's best not to think about how they know that). It varies in thickness, too – it's about 0.55mm (¹⁄₁₀₀in) thick on your eyelids, and well over 1cm (⅓in) on the soles of the feet. Without this protective layer, we'd all be doomed – it helps to keep the water in our tissues in, and helps to protect us from dangerous bacteria.

(Not that you were planning on removing your skin, but it's not something I'd recommend.)

Look at your skin very close up **(don't stick your finger in your eye)** and you'll be able to see that it's made of a thin, outer layer called the epidermis. As skin cells multiply, they make their way to the body's surface, where they die and create what looks, under a microscope, like a horny, protective layer. It's odd then that your skeleton is what gets the bad rap, when your outer layer of skin is what's really dead!

This epidermis also contains cells that contribute a pigment, melanin, that filters out damaging ultraviolet waves contained in sunlight – and it's this that affects what skin colour we have. Below the epidermis lies the dermis, which has elastic fibres running through it, so that the skin can bounce back into place if it's stretched. Then go one more level down and you'll find the hypodermis, which is a layer of fatty tissue which anchors the skin to the muscles underneath and basically stops your skin from falling off!

5. THE LYMPHATIC SYSTEM
AKA the body's police force

If white blood cells are like the blood's police officers, then the lymphatic system is like the entire body's police force. Much like your blood system, you'll find lymph vessels weaving their way throughout your body, but rather than carrying blood, these tubes carry... lymph! This is a colourless liquid full of white blood cells that fight disease by attacking any bacteria and viruses that might upset the body's workings.

Bacteria, by the way, are living cells, while viruses are much smaller – they're bits and pieces of genetic material housed within a protective coating. In fact, they're so simple, that scientists can't even agree if they should be classed as alive or not!

So where does this lymph fluid come from? Well... this is the part where it's useful to think of the lymphatic system less like the police and more like a drainage system. (Although it's still sort of like a police force as they are at least cleaning up crime!)

(EDITOR'S NOTE) ENOUGH WITH THE TIME MIXED METAPHORS, CHARLIE.

As your blood flows around the body, fluid passes from the blood vessels into your body tissues where it then carries nutrients to the cells. But it also picks up any waste (known as 'pathogens') which are damaged cells, general waste products or harmful bacteria, so that these things don't get in the way of your body's processes. All of this waste then drains into the lymph vessels and ends up in one of your hundreds of lymph glands.

These lymph glands look like little bean-shaped nodes – in fact, you may sometimes be able to feel the ones in your neck and in your armpits, particularly when they swell up as your body fights off an infection. If you've ever heard someone moan, 'My glands are up!' then that's what they're talking about. These lymph nodes are packed with more white blood cells that filter the lymph fluid so it's cleaned of nasty stuff and ready to be returned to the bloodstream.

There are various different organs that come under the lymphatic system, the biggest of which is the spleen, which is a bit like a giant lymph node. Unlike, say, the heart, the spleen is one of your organs that you can actually live without, as other body parts overlap its functions – however, losing it would make you more prone to infections.

FUN FACT: Subconsciously, we tend to dislike the smell of people who have a similar immune system to us, which means we're more likely to choose a mate with a very different one, which would give any potential children the benefits of both in their DNA. Researchers have tested this by letting people choose their favourite sweaty T-shirt. Eau de gym, anyone?

6. THE RESPIRATORY SYSTEM
AKA just breathe, man

When you take a big breath, this is the system that you're using – although while the breathing itself is obviously pretty crucial, there's a lot more to it than just the ins and the outs. In order to breathe in the air, take in the useful oxygen and eventually breathe out the toxic carbon dioxide, many different processes have to happen in between.

Breathing is controlled by the brain, although this happens without you having to think about it – the brain regularly sends nerve impulses (which I'll explain in a bit) to the muscles around your lungs. These impulses make your ribcage lift and moves your diaphragm downwards – this being a dome-shaped muscle under the lungs. It's these muscles that give your lungs the space to expand, which in turn changes the air pressure and causes air to rush into the lungs through your nose and mouth, and down through your airways.

Your lungs are perfectly designed to get exactly what the body needs from that breath of fresh air. Each spongy pink lung is full of millions of tiny air sacs called alveoli, which give your lungs a surface area of around 100m² (1,076ft²). These air sacs are covered in loads of tiny blood vessels, and the wall between these vessels and the sacs is only one cell thick (or 0.00001cm, less than a 1,000th of an inch) so oxygen can easily pass through it into the bloodstream. This oxygen is then whisked away all around the body by red blood cells in the arteries, and then carbon dioxide comes back through the wall into the air sacs, ready to be expelled!

Then it's time to breathe out again, and so the muscles between your ribs relax, as does your diaphragm, which squeezes the insides of your chest and the air from your lungs is forced out. On average, you're breathing between 17,000 and 23,000 breaths a day. Although you obviously don't need to think about breathing, as soon as you do start thinking about it, it's hard not to control it (sorry there wasn't a way to explain that without making you think about it yourself).

FUN FACT: The average adult can hold their breath for between 30 to 60 seconds, but free divers, who plunge to depths underwater without scuba gear, train themselves to do so for much longer. In 2016, a diver from Spain held his breath underwater for an incredible 24 minutes! (Show off.)

7. THE DIGESTIVE SYSTEM
AKA you are literally what you eat

As well as providing the energy we need to move around and for our body to function, food also gives us the vital components that we need to build and repair our tissues. As I've said before, all atoms are ultimately recycled, and so our bodies are ultimately formed from the food we consume.

In order for this to work, we need to break the food we eat down into small molecules so the nutrients the food contains – carbs, protein, fats, vitamins and minerals – can be absorbed by the blood and distributed to the various parts of the body. The system that's responsible for all of this, the digestive system, is about 9m (30ft) of tubes.

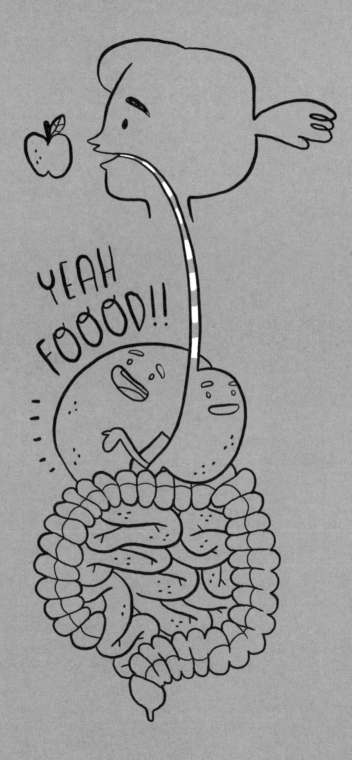

THIS ALL BEGINS IN THE MOUTH WITH CHEWING AND CHOMPING, BUT YOUR BODY DOESN'T EXPECT YOUR TEETH TO DO ALL THE HARD WORK. THE CHEMICAL BREAKDOWN PROCESS HAS ALREADY STARTED WHEN YOU PUT FOOD INTO YOUR MOUTH, AS YOUR SALIVA CONTAINS DIGESTIVE ENZYMES THAT HELP TO BREAK DOWN LARGE MOLECULES OF FOOD INTO SMALLER ONES THAT CAN BE MORE EASILY ABSORBED.

As we swallow, muscles push the food down the throat so that it eventually reaches the stomach, a muscular pouch that can pummel and churn the already well-beaten food with more digestive enzymes to break it down even further. (I assume my habit of personifying literally everything is to blame for this, but now even *I* feel bad for the food I eat!) The juices in the stomach also contain hydrochloric acid, which both kills off any unwanted bacteria and helps to dissolve your meal. (This acid is the same as toilet cleaner!)

After everything has been thoroughly digested in the stomach, what's left passes to the small intestine. This is a long, thin tube which is covered in tiny hair-like structures called microvilli – making it the perfect place for absorbing lots of nutrients and water from the digested food. In fact, by the time it reaches the large intestine, there's not much left to absorb but water. What remains is no longer useful to the body, and so it's removed in the form of... well, you know what. This isn't quite the end of the process though – your liver, which is actually the body's largest gland – acts as a tiny chemical factory and processes the nutrients absorbed by the blood into substances the body can use.

8. THE NERVOUS SYSTEM
AKA the HQ Centre

This system is, in many ways, the HQ of it all – the centre of all your mental activity, every single thought. It includes the brain, the spinal cord and the nerves, which are fibres that transmit information from all parts of your body to the brain and spinal cord, and send nerve impulses – kind of like signals – back to your organs and muscles.

Your body has millions of specialized sensory receptors that pick up any changes going on outside of it – in temperature, light, sound, pain (which technically is the feeling you get when something is damaging your tissues), pressure and so on. These receptors might be gathered together in a specialized sense organ – like the eye or tongue – and some bits of you are, of course, more sensitive than others. **(Hey-ho!)**

I LOOK SO CREEPY!

FUN FACT: Did you know that you actually have nine senses, not five? As well as sight, hearing, touch, taste and smell, you also have senses of heat, balance/pressure and pain. The final one is known as proprioception (probably my favourite sense) which is the sense of where your body parts are in space!

YOUR NERVE CELLS, WHEN ALL LINKED TOGETHER, TRANSMIT THIS TSUNAMI OF INCOMING INFORMATION AS NERVE IMPULSES, WHICH ARE ESSENTIALLY ELECTRICAL IMPULSES WHIZZING AROUND YOUR BODY! IT'S A BIT MORE COMPLICATED THAN YOU JUST FUNCTIONING AS A BIG POWER CABLE, THOUGH: YOUR NERVE CELLS, OR NEURONS AS THEY'RE CALLED, DON'T ACTUALLY TOUCH, SO THEY SEND SPECIAL MOLECULES — CHEMICAL MESSENGERS — ACROSS THE TINY GAP (THE SYNAPSE) BETWEEN THEM TO MAINTAIN THE NERVE IMPULSE.

As well as these neurons, glial cells are also crucial to the nervous system. Their name comes from the Greek word for 'glue' (although it would have been useful if the Greeks had a word for wire insulation). These cells are super cool – or non-excitable – which means that they don't conduct electricity, and thus can act as a supportive tissue for the nervous system.

Whenever new information reaches the brain, it'll respond in turn by telling muscles to contract, or glands to secrete substances – whatever it needs the body to do, basically. Only some of what it's doing is you consciously making decisions – "I need to make myself a cup of tea!", for example, but a lot isn't, which is a pretty good thing. Just imagine having:

"MUST KEEP HEART BEATING!" and
"DON'T STOP BREATHING!" on your to-do list at all times?

Sometimes, though, nerve signals don't even need to travel all the way to the brain for the body to react. Instead, the spinal cord can fill in the role of the brain (thanks for stepping in, spinal cord!). While it normally carries messages up to the top, it also handles the body's reflexes – reactions to things that don't require any conscious thought, but do need speed. For example, when you touch something too hot and instantly withdraw your hand, that's the withdrawal reflex. This protects you from hurting yourself more in the time it would take you to clock that the pan's too hot, have a little think about it –

"Hmm, this pan's a little hot. Perhaps I should do something about that..." – cue removing your sizzled fingers.

Heck, even when your brain is the one doing the thinking, often your conscious mind is the last one to know about it. You know the physical sensations your body goes through when you're feeling scared... the tensing in your stomach, for example? Well, when your eyes see something you should be afraid of, your brain's first priority is actually to tell your stomach to tense up... and then your conscious mind notices that your stomach is tense, and realizes that the thing you're looking at is something to be scared of! How weird and amazing is that?

9. THE ENDOCRINE SYSTEM
AKA the EMO zone

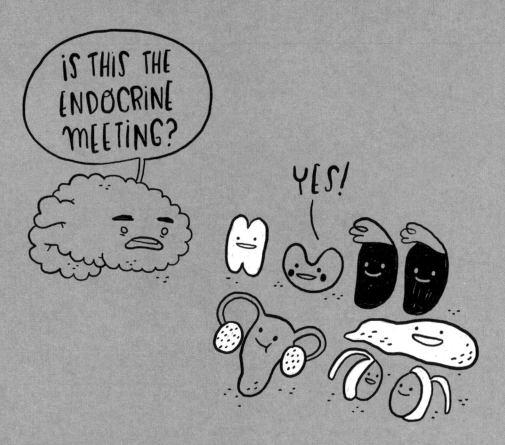

Nope, it's not just teenagers – in so many ways, you really are driven by your hormones.

The endocrine system is the one in charge, and often works together with the nervous system. The hypothalamus, an almond-sized portion of the brain, is what links the two together.

AS WELL AS ALL THOSE ELECTRICAL SIGNALS FLYING AROUND YOUR BODY, YOUR HORMONES — WHICH ARE CHEMICAL SIGNALS THAT THE BODY USES AS A WAY TO GET INFORMATION TO CELLS — CREATE MORE ACTIVITY STILL. HORMONES MOVE THROUGH THE BLOODSTREAM, SO THEY ACT MORE SLOWLY THAN NERVE IMPULSES, AND THEY AFFECT PROCESSES THAT TAKE A BIT OF TIME TOO — LIKE CELL GROWTH.

(Basically, if your hormones were the ones sending the good old... **'DON'T STOP BREATHING'** messages, then you'd be dead pretty quickly.)

Your body's hormones are produced by a network of glands, which are little organs that store and secrete them when necessary. For example, your adrenal glands, located on the top of your kidneys, make adrenalin – the **'fight or flight'** chemical that in times of stress makes your heart race and rushes blood to your muscles and brain, prepping you to punch someone or do a runner. It's very handy if you're being chased by a bear, but given the relative safety of modern life, these days it's more often secreted when you're fretting about an interview or how to write a book on science (so, not quite as useful).

But perhaps one of the most fascinating hormones is oxytocin – often called the love hormone. Produced by the hypothalamus, this is the stuff that stimulates contractions of the uterus (womb) when a woman is having a baby, which is key to the birth process. But it's also thought to be involved in everything from regulating our sleep cycles, to playing a role in how mothers bond with their babies, to how adults fall in love, and even our orgasms.

EDITOR'S NOTE) CHARLIE THIS IS MEANT TO BE PG-RATED SCIENCE.

Basically, if there's any chemical that is vital to a love potion, it's this one.

10. THE REPRODUCTIVE SYSTEM
AKA the love zone

The reproductive system is a great illustration of how none of these bodily systems work in isolation. Hormones produced by the endocrine system are required to stimulate a woman's ovaries to release eggs, or prompt a man's testes (or testicles as you probably know them) to make millions of sperm.

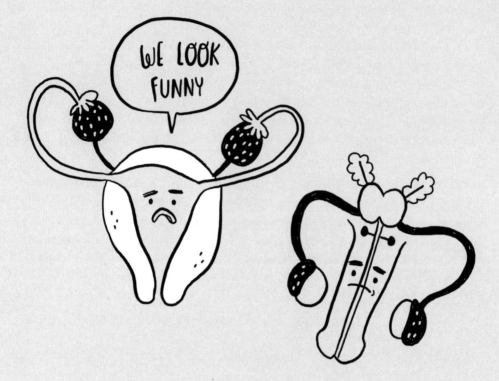

As you have probably figured out, the male and female systems are incredibly distinct (see above) – and they both function in very different ways too. For example, boys don't produce sperm until puberty – and then they start to produce new sperm continually – while a baby girl is born with up to 2 million egg cells in her ovaries in advance. Of these eggs, though, only around 400 will develop into mature eggs and go through ovulation, a monthly process that kicks in once she's reached puberty – when her reproductive system is ready to go to work.

Here's how the system works: every month an egg is released from the ovary and passes to the uterus, ready to meet a suitable sperm. During sex, sperm cells – armed with tails to propel them through the body, and with enzymes on their tips which can break down an egg's protective coating – travel through the vagina and enter the uterus. There, if just one of the many millions of sperm manages to reach the egg cell and fuse with it, you've got a fertilized egg!

This egg then divides, and then divides again, and then – well, you get the picture – over time it develops from a clutch of cells into a baby. After 9 months, the uterus's muscular walls begin to contract, pushing the baby out of its opening and through the vagina. But if no egg is fertilized, the uterus lining is just shed during a woman's period, ready for a new egg to get a chance the following month.

It's a system that, from the body's perspective, is really no different from any other... but yes, I did still feel kind of naughty writing all of that.

that's vaginas and willies FoR You I GUeSS!

11. THE URINARY (RENAL) SYSTEM
AKA drip, drip, drip

SPEAKING OF PRIVATE PARTS — LET'S TALK ABOUT URINE!

So, I mentioned before how the lungs get rid of carbon dioxide, and the digestive system expels stuff that we can't break down, but obviously that's not the only way your body gets rid of stuff it doesn't want. For this, you need your kidneys: two reddish brown organs found just below your ribcage.

Inside them, millions of tiny filters collect waste from the blood – a mixture of excess water, salts and urea, which is what's left over when your body has broken down protein. Cleverly, the brain monitors the concentration of your blood for you and controls how much water the kidneys are reabsorbing to stop the blood becoming too watery or, conversely, too thick and syrupy. Meanwhile, the collected water and waste – what we call urine – passes into the bladder where it's stored until the bladder is full... and then when finally you feel like you need to wee, you relax the muscles that keep your bladder closed, allowing the urine to pass along a tube called the urethra and out of the body. **DRIP. DRIP. DRIP.**

(I'm just trying to make you feel like you need to go to the loo... so you can try this process out yourself now that you know how it works. Let me know if it did the trick.)

To zoom out for a moment, consider all of these different systems and how they work in tandem – each really is its own triumph of evolution. The more you learn about them (and there really is a lot more to learn – I've only skimmed the surface in this chapter) the more you can see how they've all been honed over the years to solve the various problems that can arise when you're dealing with such a fragile thing as life... and to think, as well, that all these systems are made possible through the combined efforts of all of the tiny organisms that make us up – about 30 trillion individual little, living cells.

The closer you look at a person, the harder it is to comfortably call them an individual. Or as, Carl Sagan neatly put it:

"We are each of us a multitude. Within us, is a little universe."

(Carl – sorry for tacking one of my favourite quotes of yours onto to the wee section of this chapter.)

CHAPTER 6

THE

BRAIN

IF THERE'S ANY ORGAN THAT DESERVES ITS OWN CHAPTER, IT'S THIS ONE; THE CONTROL CENTRE OF THE BODY. ALTHOUGH... GIVEN THE DUBIOUS NATURE OF FREE WILL, MAYBE THIS IS JUST WHAT MY BRAIN WANTS ME TO WRITE? LET'S WORRY ABOUT THAT LATER...

THE HISTORY OF THE BRAIN

For Earth's earliest inhabitants, there wasn't much need for a brain. After all, when you're just a single cell, chilling out in the Earth's ancient oceans... there's not exactly a lot going on that requires the need for pondering.

Many years later, the next creatures that emerged on the scene were very simple animals made up of just a few cells which were also perfectly content spending their days brainless. However, in order for these cells to function together as one single organism (these cells were a team, after all!) they somehow needed to find a way to talk to one another. Thus, using a mix of chemical signals and electrical pulses, the cells were able to chat – and these early methods for communication became the basis of the processes by which the cells in your body still communicate with each other today.

Eventually, as organisms grew more complicated, some cells became particularly good at carrying these cellular messages. (SORRY, that pun was just too good not to use.)

These cells were the neurons (or nerve cells) which formed something like a brain, but instead of the cells collecting in one place, they were scattered throughout the body, forming a sort of net of nerves. In fact, some animals, such as jellyfish, still rely on this kind of nervous system. However, in other creatures (probably little worm-like things) a small hub of specialized neurons began to gather together near to the mouth, along with some rudimentary eyes. This humble little organ was how your brain began.

In the intervening years between then and now, the brain has grown into the incredibly impressive organ currently sitting in your skull. It's made of about 100 billion neuron cells that are constantly sending messages to each other via trillions of neural connections called synapses. Right now, there are many thousands of these electrochemical signals firing around in your brain, and the combination of all of them together create... well, you! Your thoughts, your consciousness... even the processes required to take the words from this page (via your eyes, of course) and form them into something that makes sense to you **(I hope)** – and it's all just thanks to those tiny connections in your head.

The neurons in your brain aren't alone, though; in fact they're far outnumbered by what are known as glial cells. For a long time these were thought to do nothing more than support the nerve cells (i.e. keeping them in place, insulating their electric impulses)... but they might actually be hiding much more exciting secrets. A specific type of glial cell known as an astrocyte tend to show up in greater numbers in animals with more complex brains, and they seem to send information to neurons too, suggesting they're also involved in the process of thought. The brain, it seems, still has some mysteries hidden inside it...

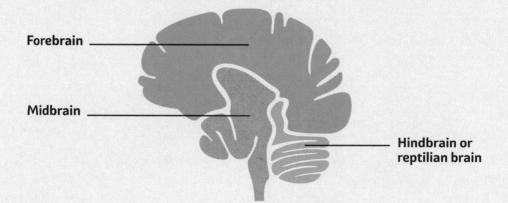

Forebrain _____

Midbrain _____

_____ Hindbrain or reptilian brain

The whole brain is made up of separate, specialized areas, each with their own functions – but all of them work together too. At the back of the brain you'll find the aptly-named hindbrain which also includes the upper part of the spinal cord. In evolutionary terms, the hindbrain is thought to be the oldest part of the brain – so old in fact that it's also referred to as the reptilian brain. So, for the reptiles living on early Earth, this hindbrain was all the brain they would have had – but it was all they actually needed! It handles automatic behaviours, such as your heart rate and breathing, as well as your fight-or-flight response when you're faced with a threat... pretty much all the basic stuff that we, animals, need to stay alive.

Because this part of your brain is so ancient, it's also the bit you can blame for some of your more animalistic behaviours. Procrastination, for example **(something I've experienced a lot while writing this book)** is when you put off a task that you know will earn you long-term rewards, and instead do something that will give you short-term satisfaction... (like watching TV). Your limbic system is what's causing this, as it's not really smart enough to know that taxes, for example, are important, even if they're not fun – so when you play video games, you're basically entering into a battle with the prehistoric, lizard part of yourself.

193

Meanwhile, the midbrain – which is in, you guessed it, the middle of the brain – is key to another automatic process. It's constantly using all of the information that's coming in from your eyes and ears and adjusts your muscle movements accordingly – constantly tweaking them so you maintain your balance, and don't fall over! And it does all of this without you even having to think about it. **How very considerate.**

Lastly, then, you have the forebrain. This includes the cerebrum – the topmost and biggest part of the brain – which in humans has grown so big that it actually covers everything else. When you think of what a brain looks like, this is the part you'd actually see – the outer, wrinkled grey matter called the cerebral cortex (cerebrum). So, when people talk about grey matter, it's this outer part that they're referring too.

Also, when people talk about the left and right brain, the cerebrum is the part that they're referring to. It's split into two halves, or hemispheres, which are separated by a deep groove but can still communicate with each other using nerve fibres at the bottom of it. Each of these hemispheres deals with different responsibilities – the left seems more focused on handling language and words, while the right deals with spatial awareness and helping you to make sense of what you see.

Clearly, it's such a complex organ and we still don't entirely know how it all works – so how exactly did we go from those mindless cells in the ancient sea all the way to, **"I think, therefore I am?"**

Well, once again, we have evolution to thank. It's pretty simple: the bigger the brain in relation to the size of your body, the smarter you are, and the more likely you are to survive! So, over millions of years as different species evolved, brains began to grow in complexity. For instance, they developed structures that could remember how good it felt to find some lunch, while also remembering how almost becoming someone else's lunch didn't feel that great at all – these were the basis of early emotions. However, not all brains were created equal: different parts of the brains got bigger depending on what helped their owners survive. So, in early mammals, the olfactory bulb (the part of the brain which is responsible for smell) swelled, which helped them sniff out food much better.

Similarly, members of my family have more Earl Grey matter, which helps us to be more discerning tea drinkers. (OK, not really. But a guy can dream.)

One of the most important brainy developments (at least for us, humans) seems to have been the huge growth spurt that happened in the neocortex of our primate ancestors. This is the wrinkly outer layer that covers the brain, and it's the biggest part of the cortex... although it didn't grow so much in thickness as it did in surface area. This is why the brain sort of looks like a walnut – those dense folds make the surface of the brain as big as possible, which allows for more complex processes.

Truly, if there's any part of the brain that separates us from the other animals, it's our cortex. It seems to be key to our uniquely human way of thinking – handling certain abilities like imagination, language, and abstract thought. Basically, this means that we're not just focusing on what's in front of our eyes, but also thinking about the big questions, like... **"why does Kim Kardashian have such a large gluteus maximus (heiney)?"** So why did we end up with such a big one? (Cortex, that is.) Well, it might have been because our ancestors lived in groups and, sort of like today's teenagers, they spent huge amounts of brainpower keeping track of their increasingly complicated social lives – something that a nice, big neocortex helped them to do.

As the early human brain continued to evolve, the various parts of it were becoming better connected – just like any tool, when it comes to brainpower, it's not just about what you've got but also how you use it. This meant that our ancestors' brains were able to cope better with incoming information, and make smarter decisions in response to it.

That said... as far as brains are concerned, bigger is always better. In the animal world, there is a direct link between brain size and intelligence. There's no need for us to be shy about our size though, as the human brain is disproportionately large compared to our body.

So, is there a limit to how giant our noggins can get? Well, the slightly awkward reality is that our brains actually have stopped growing – in fact they haven't really changed much in size for the last 200,000 years. So what happened? Well, a big brain means a big head, which means a tight squeeze when you're entering into the world out of your mother. It seems that over a certain size, the process of birthing such big-headed babies just becomes too fraught with danger for mothers and babies and, by extension, the species. This is evolution, people – there's always a give and take.

Actually, in the past 10,000 to 15,000 years, the average size of the human brain (compared with the body) has actually shrunk by about 4%. So why is that? Maybe our brains are just slowly becoming more efficient, and are now able to pack more activity into a smaller space? Well, it's also possible that we really could just be getting stupider because in today's society being able to outsmart a rival isn't typically as life-threatening as it once was. So long as you can survive long enough to get a mate and pass on your genes, it doesn't really matter if you're an Einstein or a Donald Trump – your intellect is in the genepool regardless.

Regardless of what direction our brains might be going in, though, it's still the most important organ you've got – and your body knows that too, which is why it's done such a stellar job of keeping it protected.

The first and sturdiest barrier is, of course, the skull, which surrounds your brain and acts as a protective shield. Then there are three membranes surrounding the brain, each protecting it from the skull (like a helmet). The brain itself is swilling about in a liquid called cerebrospinal fluid which also helps to cushion it from any knocks. Give your head a little shake, and you might be able to feel it swirling about in there! Or maybe that's just me...

Your brain is even slightly protected from your own blood – there's a barrier between it and the walls of your capillaries (the smallest blood vessels) which stops as much stuff getting through as they do elsewhere in the body. Oxygen can pass through this blood-brain barrier, but larger molecules, such as glucose, have to be ferried across the border by special proteins. Its purpose is to keep harmful bacteria from entering the brain... but it won't protect you from everything. This barrier does allow for things like caffeine, alcohol and nicotine to slip through. So, if you do go a bit crazy on a night out, you can always blame your brain-blood barrier for not doing a good enough job!

FUN FACT: Men and women's brains really aren't all that different. In fact, it's very difficult to lay down hard and fast rules about a typical male or female brain. One difference is that men tend to have a larger amygdala, which is an area typically associated with emotion! So a lot of our stereotypical ideas about how men and women think could be called into question by studying what differences there actually are.

PERCEPTION: THE MIND'S EYE

So, you've got this big, beautiful, well-protected brain, which is constantly receiving a torrent of incoming information from the nerves all over your body. How, then, does this wonderful organ turn all those electrical impulses into the 3D, multi-sensory picture of the world around you?

Well, the main thing to realize is that the world as you understand it (while entirely real) is the function of the brain – it's basically a vision that your mind makes.

In a way... **IT'S ALL IN YOUR HEAD!** (But it's also definitely there.)

Sight, for example, involves light hitting a part of the eye called the retina; smell means that certain molecules have reached receptors in your nose; hearing is waves of pressure – sound – being registered by the delicate apparatus in your ear. So it's all real stuff that we're measuring... but that's only half of the story. Your brain is processing all this mess of information and trying to make sense of it all – it's recognizing some things, paying particular attention to others, and sometimes even making things up by drawing on its expectations to better interpret it all.

FUN FACT: While it might be the epicentre of all feeling... the brain cannot feel pain. There are no pain receptors in the brain, but only in the surrounding tissues. That's why, when people undergo brain surgery, they sometimes only need a local anaesthetic!

It's a complicated process, and the brain handles these multiple amounts of stimuli in different ways. For instance, in order for your brain to register a change in your environment, it needs to cross a specific threshold – basically, if something has only changed a little, then it's not worth the effort registering. For instance, you'd have to turn up the dimmer switch on a light bulb to increase its intensity by 8% for your brain to notice any change!

This explains why, when I work all day on writing this book you are reading, I often suddenly look up and realize that it's dark out – the change in light happened so slowly that my brain didn't think it was worth pointing it out.

In fact, (if you'll forgive the phrase) let's look at vision more closely... To us, seeing seems like something our eyes and brain are flawless at. Your brain does a brilliant job at keeping the shape of things constant, even when they are moving. So, if someone walks towards you, taking up more space in your field of vision as they get closer, your brain is smart enough to register that they aren't actually slowly becoming a giant right before your eyes – even though the image of them in your brain is growing. Similarly, when you're watching a film even though you're seeing 24 individual still images every second, your brain translates those images in your mind into seamless motion because of something called persistence of vision – your brain mentally fills in the gaps between the frames!

All of this is just your brain doing its best to make sense of the world. This stuff kicks in when we're really young too – take depth perception, without which the world would appear to be 2D. Scientists have actually carried out experiments recreating what would happen to a baby on the edge of a cliff... by putting a baby on one side of a glass-topped table to give it the impression it's perched on a precipice. **(Scientists aren't monsters, I promise!)** Babies will refuse to crawl off the edge in front of them (really, just onto the glass surface), signalling that they can already perceive the depth in front of them. Clever babies!

However, perception definitely isn't perfect
– in fact, sometimes it's downright confusing.
It's heavily influenced by what you're expecting
to see. Have you ever seen a face in an
inanimate object, for example? Well, that's
called pareidolia, and it happens because your
mind is wired to recognize familiar human faces.
**(Perhaps why some people are so insistent that
they've seen the face of Jesus in their toast.)**

But our brain glitches don't end there, as it
can even mess with your perception of time!
Intense fear, for example, can result in the
apparent slowing down of time, with survivors
of dangerous situations reporting that things
seemed to take much longer to happen. As
it turns out, fear doesn't actually speed up
your rate of perception but it does help you to
remember events in more detail – and because
you remember more things from a set period of
time, such an experience feels like it unfolded
slower. The same thing happens in reverse,
too – if you spend your entire day staring at a
computer then your mind won't be registering
any new experiences, and so it'll seem like no
time has passed at all!

MOTOR CONTROL

Even when you might be too engrossed in something to notice what's going on around you *(for example let's say you're reading an awesome science book)* your brain is clever enough to keep tabs on your surroundings without you having to think about it. It's always ready to act based on all of the information it's taking in – it does this by coordinating all the muscles involved in movement to make an appropriate response, be that running for your life or helping you to take those few extra steps to fall into bed (and not onto the floor) at the end of a night out.

We call this motor control. A lot of it's done automatically: you don't have to think about putting one foot in front of the other when you walk *(unless you're very drunk)* and, equally, a lot of the changes you make to your posture during movement are done unconsciously.

While reflex movements don't involve the brain – as I mentioned before, reflexes need to be timely and so the spinal cord handles them instead – the brain is still the boss in terms of all of the body's voluntary movements. The key area of the brain that deals with this stuff is the motor cortex, located in the brain's frontal lobe. Weirdly, signals from one side of the motor cortex control movement on the opposite side of the body:

So, if you start waving your right hand around (try it now! I've made you do weirder things) then it's actually the left side of your brain that's operating it.

In fact, every part of the body corresponds with a specific area of the motor cortex. Rather gruesomely, scientists found this out in 1870 by applying electricity to different parts of a dog's motor cortex, and watched what part of its body started twitching. Later researchers mapped out the human motor cortex in a similar way – by stimulating parts of a patient's brain with a tiny electric current and seeing what part of the body started tingling or, in some cases, moving.

FUN FACT: There's no such thing as multitasking! It does not exist, in your head, at least. After scanning the brains of people who were trying to do multiple things at once, it was revealed that all they were really doing was switching back and forth between tasks and actually just doing one at a time, and they were even spending excess brainpower every time they switched between them.

Naturally, the biggest areas of the motor cortex control the most complicated parts of the body: the hands and face. This is simply because of the huge and varied amount of movements you can make with them, from manipulating tools to producing different types of sounds as part of speech. However, use of your motor cortex isn't fixed – it can actually change depending on what you need the most. For example, among blind people who read Braille using one index finger, a large part of their motor cortex can become devoted to that single finger! Cue everyone wondering how they could give certain other body parts slightly more room in the motor cortex...

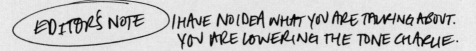

EDITOR'S NOTE) I HAVE NO IDEA WHAT YOU ARE TALKING ABOUT. YOU ARE LOWERING THE TONE CHARLIE.

Some scientists have really gone all out in their investigations into how the brain controls movement. Jose Delgado, who pioneered electrical stimulation of the brain, went as far as developing a radio device which allowed him to stimulate a brain remotely. In one famous demonstration, he let a bull charge at him and simply by pushing a button, stopped it in its tracks! Stimulating the bull's brain had probably made its brain activity confused, leading it to come to a halt. Imagine if all science experiments were that show-stopping!

While we might slowly be learning how these motor controls work, even to the point where we can manipulate them ourselves, we're still ultimately at the whim of our own brains.

Here's another test to try out: Sit down, raise your right leg, and start turning your foot in clockwise circles. Then, trace the number six in the air with your right hand...

Notice anything odd? If it worked, then your foot is now turning in the other direction, without any conscious decision by you to do so! Which leads us to a really important question...

JUST WHO IS IN CHARGE HERE?

CONSCIOUSNESS AND FREE WILL

WHO AM I?*

*Don't worry, dear reader – I haven't lost my mind just yet. Instead, I'm posing that question so that you can ask it to yourself. Go on, who are you?

And thus, by thinking that thought, you're exhibiting consciousness – an awareness of your own existence as an individual. Your inner monologue, your decisions, your ideas... it's weird to think that the things that make us **Us** all exist within certain sections of this squidgy blob.

What's the use of being an *I*, anyway? Other animals seem to get on just fine without it. Well, the answer could be that for our primate ancestors, having an awareness of being an *I* among a lot of other *yous* – each with their own feelings and plans – was handy for navigating the complexities of the early social world. After all, it's good to know who's the most likely to share some fruit with you. Then, once you were able to appreciate that other creatures had a mind of their own, you could recognize that you had one too! This was the beginnings of consciousness...

Anyway, that's one idea – there's not total agreement on this topic just yet. Some people think that it's by experiencing sensations, not social interactions, that you first gain a sense of self. Or perhaps internal signals, such as hunger, thirst and pain, are the first prompts towards consciousness. However it happened, the burden of consciousness is ours to bear as humans, and so bear it we shall. **(Mostly by watching a lot of bad TV and worrying a lot. I think it's what our ancient ancestors would have wanted.)**

So, when you're experiencing conscious thought, what exactly is happening in your head? Well, the answer to that question is probably another Noble Prize waiting to happen – because as of right now, we're just not sure. Although, in very simple terms, it seems that various parts of the brain have to communicate with each other, and neurons in the front of the cortex (associated with emotions and decision-making) have to get busy in order for consciousness to pop up.

By contrast, when one of your behaviours is happening automatically, without your being conscious of doing it, the activity doesn't seem to resonate through the whole brain in the same way. Tests have shown that if a word is flashed before someone's eyes so quickly that the person doesn't report seeing the word, only the visual bit of the cortex gets busy – in this case, your brain has seen the word, but **you** haven't.

ARLE IS AMAZING CHARLE IS AMAZING CHARLE IS AM
CHARLE IS AMAZING CHARLE IS AMAZING CHARLE IS
E IS AMAZING CHARLE IS AMAZING CHARLE IS AMAZIN

There is still a lot of mystery around consciousness. What this suggests is that we're taking in a lot – registering the stuff around us, noting emotional signals from people – without ever realizing that we're doing so.

But wait, there's more...! While we very much like to think that we can control our bodies ... is that really how it works, or is our brain in control of us? This is not an abstract, philosophical question though, people, it actually was tackled by an experiment done in the 1960s. In this experiment, the subjects' brainwaves were measured and they were told simply to wiggle their finger whenever they wanted. Now, what you'd expect to see is a blip of brain activity just before you wiggle your finger, signifying your decision to move it... but what was discovered is that brain activity is seen a second before you make your decision. A whole second! That's a clear indication that your brain is ahead of you – almost as if your brain is willing you to will. Does this mean that free will is an illusion? Well, this test definitely does make it seem that way...

LEARNING AND MEMORY

When it comes to your memory, the magic number is seven (well, somewhere between five and nine, actually). As this is the amount of items most adults can store in their short-term memory, as shown by psychological experiments.

The definition of short-term memory is when you only need to keep something in your head for just a moment or two (up to about 30 seconds) – such as a phone number you're about to tap into your phone.

There are a few different ways that this can work – for instance, when you try and remember a phone number you've just heard, do you see it flash up in your mind's eye, or do you hear it in your head? Well, it seems we mostly rely on the latter, known as acoustic encoding. So, if you think you're more likely to do the former, which is visual encoding, then congrats for breaking the short-term memory status quo!

So, what happens when you're remembering, say, your first kiss? How does that differ? Well, when you're thinking about short-term memories, neurons in the area right at the front of your brain – the prefrontal lobe – get very busy. In contrast, long-term memories aren't just a temporary flurry of activity, but they've actually altered your brain. Long-term memory might represent something you remember for just a few minutes or a whole lifetime, and covers everything from general knowledge to how to brush your teeth. No matter, when a long-term memory is laid down, neurons make new physical connections with each other – which stick around whether you are accessing the memory or not. It's the hippocampus, a structure deep within the brain, that encodes incoming information into a long-term memory by forming these new connections. When you want to remember an experience, the hippocampus also helps you to play it back to yourself by activating the parts of your brain that deal with smell, taste, sound and vision, so you actually recreate the events in your mind. But the hippocampus has so much information to deal with, it can get selective – making memories more effectively if something carries an emotional punch, or if it is getting a lot of attention from your short-term memory. So why do we remember some kisses more then others? That might depend on how interested you were in the kiss, or how much you enjoyed it. (Cheeky.)

However, don't think of these long-term connections like data stored in a hard drive or like books locked up in a library. This is still biology and memories definitely get altered slightly over time. In fact, because your brain recreates a memory when you recall it, every time you think of that event, there's a chance it'll be recreated slightly wrong.

SO, THE MORE YOU REMEMBER SOMETHING, THE MORE LIKELY IT IS THAT YOU'RE NOT REMEMBERING THAT EVENT AS IT HAPPENED AT ALL! IF ANYTHING, THE SAFEST MEMORIES ARE THE ONES YOU NEVER THINK ABOUT...

FUN FACT: London taxi drivers are known for their in-depth knowledge of the streets of the capital, but how has this information changed their brains? Studies show that the posterior region of the hippocampus – the area that's involved in spatial memory – is bigger in taxi drivers compared to other people. What's more, the longer they've done the job, the bigger it is too. It seems the taxi drivers' work alters their brains so they are better able to store a mental map – demonstrating how, even as adults, we're equipped to adapt to new challenges.

WHERE YOU OFF TO, MATE?

CHAPTER SEVEN, KIND SIR!

CHAPTER 7

THE CELL

THINGS ARE ABOUT TO GET MICROSCOPIC, BABY!

TYPES OF CELLS

Every living organism that you see on this planet owes their livelihood to the HUGE collaborative effort that goes on inside their bodies (including yours and mine).

I'm talking, of course, about our cells: the building blocks of life. It's not all about bodies, though – some cells can function just fine on their own. Single-celled bacteria, for example, are made up of just one cell, but still very much count as living organisms in their own right (thank you very much).

But you knew all of that already, didn't you? The real question we are about to tackle is: what even are these little things called cells...? (Was the chapter title a giveaway?)

The first thing to know is that there are two basic types of cells: **prokaryotes** and **eukaryotes**, which differ based on how they arrange their DNA inside them.

The first lot, the prokaryotes, just sort of keep their DNA wherever they feel like, letting it float around inside them in their cytoplasm (which is a gel-like fluid you find inside cells). They're also the older type of cell – bacteria, for example, are prokaryotes which have been knocking around on Earth for billions of years.

The second group, eukaryotes, keep their DNA in a neat little membrane-bound sac called a nucleus. You can think of the nucleus as sort of like the cell's brain – it controls division, which is how cells reproduce themselves. These also happen to be the kind of cells that make up us, as well as all other animals and plants! (Which personally makes me happy to know, because I like the idea of my cells being nice and organized.)

Ready for some more terrible pun-based memory aids?

I'M A PRO-KARYOTES AT KEEPING MY DNA WHEREVER I GOSH DARN LIKE! I'M TELLING EU-KARYOTES TO PUT YOUR MESSY DNA IN ONE PLACE!

So, living organisms can either be single-celled – like bacteria – or multicellular, which is what we humans are. When you have a more complex organism (like us, because we're just **so** complicated) then you'll tend to find cells that are more specialized, each with their own roles and functions. Similar cells might team up to form a tissue – lungs or stomach tissue – and make up part of an organ. That's all we are, really, deep down – just, many, many, many, **MANY** cells all working together!

Remember this teaming up point, by the way. That's going to become pretty important later... >>>

Anyway, while a single cell might seem relatively simple on the surface, there's actually a lot going on inside of them. Go inside a cell and you'll find different types of useful structures within them, called organelles – a bit like organs in a body. (But much more fashionable. These aren't organs, my dear, they're organelles.)

FUN FACT: While individual cells are usually microscopic, far too small to be seen with the naked eye, one exception is *Caulerpa taxifolia* (a terribly boring name, agreed) – an aquatic alga which is the biggest single-celled organism in the world. You'd never guess by looking at it (as it looks like a regular underwater plant) but each alga is a single cell that can grow up to 30cm (12in) long. Big shout out to the cells that were brave enough to go solo!

THE ANATOMY OF A CELL

While plant cells and animal cells do have a lot in common (they're both eukaryotes) there are some key differences between the two... one makes up plants... and the other makes up animals! **(WHOAH, CHARLIE DON'T PUSH US TOO HARD NOW!)**

OK, there's a little bit more to it than just that...

Plant cells have a strong, protective wall; they tend to be more regularly shaped (often rectangular); and they can be much bigger too – they range up to 100 micrometres (which is a tenth of a millimetre), and the biggest animal cells are only 30 micrometres. Basically, compared to animal cells, plant cells are stronger, bigger, and in better shape. **(Plants, eh? What a bunch of weeds...)**

But if that wasn't enough, plant cells have yet another impressive feature: chloroplasts, which are one of their organelles. These little structures are full of a green chemical called chlorophyll, which is the stuff that makes plants... well, green! By harnessing the Sun's light energy – which is absorbed by the chlorophyll – chloroplasts can suck up water from the plant's surroundings and carbon dioxide from the air and convert them into glucose (sugar) – fuel for the plant to grow. The process produces oxygen as a waste product, which is handy for us as it's what we need to breathe!

(OK, plants. I'll forgive your cells' superiority – but ONLY because you help to keep me alive.)

Just a thought – I guess this all means that if Superman really was able to harness the power of the Sun like he does in the comics... he'd probably need to be bright green, like a plant.

While animal cells don't have chloroplasts, they do have a few tricks up their squidgy sleeves. For example, they have odd little structures called mitochondria which take the food we've consumed and use it to produce the majority of the cell's energy. Also, while animal cells might not have a tough cell wall, their flexible, porous membrane still does a fine job of keeping the cell's bits and bobs together, as well as managing what comes in and out of it. The hormone insulin, for example, actually binds to receptors on the cell membrane to tell the cell that it's OK to take in glucose, which helps maintain blood sugar at the right level. It is sort of like the cell's own version of border control!

SORRY, THERE'S A HIGH BLOOD SUGAR ALERT RIGHT NOW.

Not all cells are born equal, though – while some are just the funny little blobs you're used to seeing, others have little appendages to help them move either themselves or other things around. The sperm cell for example has a flagellum, the little tail it uses to swim to the egg, and the cells in your airways use their cilia (tiny hairs) to move mucus (and whatever might get stuck in it) away from your lungs.

FUN FACT: Even though red blood cells are eukaryotes, they don't have a nucleus! There's always one, isn't there? Weirdly, they do actually start with all the little organelles that other eukaryotic cells have, like a nucleus and mitochondria, but they lose them as they age. Why? Well, so they can have more space to pack in haemoglobin, which is the molecule that enables red blood cells to ferry oxygen around the body! Sacrificing their organelles for the good of the body? How selfless!

Stem cells deserve a special mention, too, as they have the unique ability to transform themselves... into different types of cells! When a stem cell divides, the new cell could stay a stem cell or it could become a specialized cell, such as a heart, muscle, blood or nerve cell. In adults, a stem cell usually matches the tissue where it is found so, the cells in the bone marrow usually produce different types of blood cells, replacing those that have been lost through disease or wear and tear. That's how stem cells can help maintain and repair tissues and organs in the body – and why we're so interested in their possible medicinal uses.

...And it's also why I'm so interested in their possible ability to turn me into a shapeshifter. I mean, why stop at transforming them into normal types of cells? Come on, science. I believe in you.

EDITOR'S NOTE PLEASE STOP PUTTING THINGS THAT AREN'T SCIENCE IN THIS SCIENCE BOOK, CHARLIE.

WHEN 2 CELLS BECOME 1

OK, so now we know about these two main types of cells: prokaryotes and eukaryotes. The next question is how did we end up with them?

Well, it seems likely that the prokaryotes came first, and that the eukaryotes were their evolutionary offspring... but the really strange thing here is that, while you'd expect this to have happened as a result of a random genetic mutation (like we're used to with evolution), it actually started when two prokaryotic cells... teamed up. **(I told you it was going to be important!)**

Before we get into the story of the world's first biological collaboration, let's talk about *mitochondria*. Remember that long word from earlier? Well these are sort of like the batteries for the cells – they take energy from food and convert it into forms that the cell can use. Generally, each animal cell has about 1,000–2,000 mitochondria, and on their own they look a bit like baked beans (although normally they join up in networks that branch across their cell, presumably because of their collective embarrassment about looking like baked beans). Interestingly, mitochondria also contain some of their own DNA, as opposed to the DNA contained in the cell's nucleus.

With me so far? OK, well, here's why all this is SO relevant. The idea is that mitochondria are actually the descendants of a bacterium that, billions of years ago, was very much flying solo. However, while this little bacterium was minding its own business, it was ENGULFED by another cell, probably as a snack!

Here's the crucial part, though – rather than being broken down by this hungry cell, it actually survived and stayed tucked away inside it! This actually turned out to be a really good deal for both of them. The ancient mitochondrion found itself much safer inside its new buddy (even by bacterial standards, it was quite teeny and vulnerable) and could thrive on the nutrients inside the bigger cell. Meanwhile, the bigger cell got the benefit of the chemical energy the smaller cell happened to be really good at producing – a win-win relationship; that's the basis of how our own cells rely on mitochondria today!

As the bigger cell reproduced itself, the internal bacteria were passed on too. Eventually, the two types of cell evolved into a single organism, and today mitochondria can't survive anywhere else! That's how we think eukaryotic cells came about, so in a way our bodies are the result of that ancient collaboration!

THE FIRST COLLAB! *(A little bit more impressive than doing The Tin Can Challenge, if you ask me. If you don't know what that is by the way... please, don't Google it.)*

But wait, how do we know that mitochondria were once bacteria? Well, you can find the clues when you peer inside them: mitochondria have their own cell membrane, like bacteria do; they have circular DNA, like (gasp!) bacteria; and they replicate themselves by splitting in half, just like... wait for it... BACTERIA! This doesn't even seem to be the only instance of this happening – in plants, chloroplasts seem to have bacterial origins too. All thanks to good old teamwork.

FUN FACT: You know how you get half your DNA from your mother and half from your father? Well, while that's the case with DNA in the cell nucleus, mitochondrial DNA is a bit different. After an egg is fertilized, the mitochondria in the sperm disintegrate – meaning all your mitochondrial DNA comes from your mother. Thanks, Mum!

CELL REPLACING > CELL REPLACING > CELL REPLACING > CELL REPLACING > CELL RE..

Previously, I mentioned how sex cells (such as human eggs and sperm) are produced through a process called meiosis – this shuffles the parent's DNA even before fertilization happens to help make offspring as unique as possible. However, this is only one of the ways that cells in your body can reproduce; now we're going to take a look at the other way: mitosis!

This type of reproduction has nothing to do with sex but is in fact a much simpler technique used in the body for growing tissues and replacing cells. To make it happen, all a cell has to do is split itself in half to create two copies of the original cell! OK, well there's a little more to it than just that: before it can split into two, all the chromosomes in the nucleus – the tightly wound DNA – are duplicated, and then they line up neatly in their new pairs so they'll be evenly split when the cell divides. (The parent cell also produces more of its various organelles in advance, so there's enough to go round when it splits.)

Fission, the method by which prokaryotic cells such as bacteria reproduce, is a pretty similar process: the DNA in a bacterium splits into two, and the cell divides into two daughter cells. Under the right conditions, some types of bacteria can divide every 20 minutes... which means that in only 7 hours one bacterium can generate more than 2 million bacteria! Yikes.

All of this is asexual reproduction, where each offspring is a genetic clone of its parent. Which might actually leave you wondering... how do bacteria evolve, then? How does a species change over time when it's just constantly making carbon copies of itself?

Well, these copies aren't always quite the same – just like in sexual reproduction, genetic mutations can still occur. In fact, bacteria can evolve much faster than us, simply because of how quickly they can reproduce. Additionally, while humans are forced to transfer their DNA down the genetic line, bacteria can do something called 'horizontal gene transfer' which lets them share genes among themselves! What's more, they don't even have to be the same species in order to do this! Combine all of these techniques, and you can pretty much guarantee some new weird and wonderful types of bacteria.

Oddly however, all of this horizontal gene swapping might have happened closer to home than you think, as a few human genes are shockingly similar to those found in bacteria. Why? Well, the question has been the subject of fierce debate by scientists, but some think it's because we didn't just inherit our genes from our ancestors – instead, along the way, bacteria gave us a few of their own via horizontal transfer! If this is the case, it could be **huge** in terms of how we think about evolution: for one thing, we wouldn't owe everything to our parents – we'd have to thank a few lucky bacteria for a helping hand in our genomes, too!

YOU ARE A WALKING PETRI DISH

ONE BIT OF SCIENCE THAT DEFINITELY ISN'T IN DOUBT IS THAT RIGHT NOW, YOU ARE ABSOLUTELY TEEMING WITH BACTERIA, INSIDE AND OUT. (OH, AND YOU'RE ALSO HOME TO FUNGI, VIRUSES AND MITES... BUT LET'S JUST STICK WITH BACTERIA FOR NOW.)

The latest scientific calculations estimate that we have roughly the same number of bacteria and human cells in our bodies. For example, I (a man weighing 70 kg/11 stone) am composed of about 40 trillion bacteria (most of them in my digestive system) and 30 trillion human cells. So, technically, I'm outnumbered by bacteria! But there's no need to PANIC just yet: since bacteria are far smaller than human cells, they make up only a small percentage of our total body weight, somewhere around 2%. Still, I am a little terrified.

As babies we get loaded up with a huge amount of bacteria at birth, as well as collecting some more via breastfeeding and through contact with our mothers' skin. These bacteria travel through the body and reproduce... but, really, all of this is nothing to be scared of. While we associate bacteria with diseases, not all of them are bad for us. In fact, it's quite the opposite: those bacteria picked up from our mother help prepare our system to fight off threats. And throughout our lives, bacteria seem to play a role in keeping us healthy, while those living in the digestive system help to break down difficult-to-digest food and even produce certain vitamins for us!

Our skin alone plays host to an estimated 1,000 species of bacteria, some of which form a protective shield that stops other, potentially more harmful bacteria getting a foothold. Not surprisingly, the highest concentration of skin bacteria is found in moist areas, such as your armpits, but scientists have found a greater diversity of bacteria in dry areas such as on your forearm. Bacteria are not always looking to help out; if your immune system is compromised by a cut, let's say, even common and usually well-behaved bacteria can turn into nasty invaders – but on the whole, they're not as bad as you might assume. **#NotAllBacteria**

FUN FACT: According to researchers behind something called the Belly Button Biodiversity project, we have around 2,368 different strains of bacteria in our belly button! More of a reason not to stick your finger in there... who knows what could bite it.

VIRUSES

It all started in 1886 when scientists tried to remove bacteria from a liquid using a special filter. However... there was a problem. Even when they used the filter on liquid taken from some diseased tobacco plants, they found the filtered liquid was still transferring the sickness to the healthy plants. What exactly was to blame?

A VIRUS.

Unlike bacteria (which can be good and bad for us) viruses always seem to be a menace. They can cause illness in any type of organism, from plants, humans and animals... right down to bacteria! They aren't actually cells, they're just bits of DNA – genes – encased in a protein coat. The key difference with them compared to other organisms (although, we're not sure if they really count as being alive) is that these very small particles can only be reproduced inside host cells that they've managed to get into.

So, how do they do this? Well, first they lose their protein coating, and then they hijack a cell's own systems so that it stops looking after itself, and starts making more bits of virus. Then, once a virus has got busy reproducing itself inside a cell, the resulting viruses will eventually burst out (sometimes killing the cell) and it will take over more cells! It might sound like science fiction, but every time you get a cold, this is what's happening inside your body.

Of course, if the owner of the cells can't fend off the virus with their immune response, the organism gets sick – and diseases caused by viruses can be much worse than just a sniffy nose, resulting in anything from chicken pox to HIV/AIDS. Fortunately, outside of the body, viruses are pretty useless. Because of their odd existence, for a while they were thought to be biological chemicals rather than actual living things. As we learn more about them however, we can't be so certain...

Close the curtains and dim the lights... it's time for a SCARY VIRUS STORY!

The year was 1992, and in the city of Bradford in the north of England... something very strange had been discovered in a water-cooling tower. After a recent outbreak of pneumonia, scientists were on the hunt for the bacteria that might have been making everyone sick... but instead, they found...

SOMETHING ELSE.

Inside the cooling tower was an amoeba (a kind of single-celled organism), but inside the amoeba, scientists discovered something they thought must be bacteria – it was so complex! How could it have been anything else? Baffled, they stuck it in a freezer... and just left it there.

Many years later, the freezer was opened once more **(NO! DON'T OPEN THE FREEZER!)** and someone decided to take a closer look at this 'bacteria...' and discovered something jaw dropping. It was, in fact, a **VIRUS!** A **REALLY BIG** virus, bigger than even some bacteria, and containing more than 900 genes – much more genetic material than anyone could have expected. So, it was named the Mimivirus, with the 'mimi' for mimicking, because it looked like a bacterium. **(OK, that 'mimi' name does kind of undermine its scariness.)**

WANT TO HEAR A SPOOKY TALE?

But the tale does not end there! Since then, even more of these giant viruses have been found, including one called Mamavirus – because it was even bigger than Mimivirus. (What's with the cute names guys!?) And where was the Mamavirus found? Inside a cooling tower, of course! (Turns out these towers just make great homes for big viruses.) In fact, the Mamavirus even had a smaller virus hanging around with it, which was named Sputnik, after the first manmade satellite.

However, Mama and Sputnik were definitely not a parent and child team – in fact, Sputnik was actually behaving like a parasite on Mamavirus, hijacking her replication factory in order to reproduce, while Mamavirus made fewer and poorer quality copies of itself! In other words, a virus was making another virus sick! Talk about karma...

FUN FACT: While it might seem straight out of a science fiction film, this really happened: in 2014, scientists announced that they had managed to thaw out a giant virus they'd found in 30,000-year-old Siberian ice... and it was still infectious! Fortunately it wasn't a threat to us (this virus was one that infects amoebae) but who knows what else could thaw out as the Earth's ice melts?

CHAPTER 8

THE

ELEMENTS

ATOMS ARE THE TINY BUILDING BLOCKS THAT MAKE UP ABSOLUTELY EVERYTHING WE SEE AROUND US. IF IT'S MATTER — WHICH IS ANYTHING WITH MASS, AS OPPOSED TO THAT ILLUSIVE DARK MATTER (WHICH IN THIS CASE DOESN'T MATTER) — THEN IT'S MADE UP OF ATOMS. AND WHILE THEY MAY NOT BE THE SMALLEST THING IN THE UNIVERSE, THEY ARE STILL ABSOLUTELY MINUSCULE. SMALLER THAN A SPECK! DOTTIER THAN A DOT! TEENIER THAN A TINY... YOU PROBABLY GET THE PICTURE.

An element is simply matter that is made from just one type of atom. There are 118 of these that we know of, and if you combine these elements then you're able to make... well, everything else! Like, literally all matter. (And yes, that includes you and me.)

All you need to do is create a chemical reaction, which will allow you to rearrange these atoms in loads of different ways to create chemical compounds. For example, you could encourage two atoms to combine with a bit of heat energy – if you burn coal it'll react with the oxygen in the air to form carbon dioxide. Sometimes things want to combine anyway and will actually create energy, like when you drop sodium in water it'll start to sizzle, forming sodium hydroxide and hydrogen. Then, of course, there are the unresponsive atoms, such as helium, which would prefer not to react with anything thank you very much.

The neat thing about these chemical compounds is that they combine elements to make something completely new – tasty salt, for example, is made when a metal and a poisonous gas combine, two things we'd never dream to put on our chips.

However, we're not going to worry about things getting too reactive here. Let's keep things simple by looking at the elements.

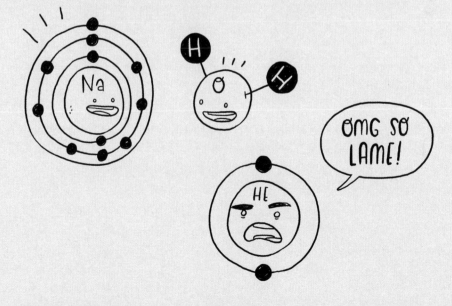

IT'S ELEMENTARY

Back in the day, the Greek philosopher Empedocles decided that there were just four elements – earth, air, fire and water – out of which everything was made. Around 350 BC, the more famous Aristotle added one more to the mix: aether, which was supposed to fill in some gaps that the other four couldn't – like how gravity and light work, for example.

These ideas were sort of on the right track, but however revolutionary they might have seemed at the time, the scientific progress we've made in the intervening 2,000 years has shown that it takes far more elements to make a universe. Not only that, but the 118 we do know about are just the ones we've discovered so far – who's to say how many more there might be out there!

To be honest, though, it took us an embarrassingly long time to grasp this whole concept. Until the Middle Ages, we were discovering new elements at a rate of about... oh, one every millennium. Our hunt for new elements only really started to kick into gear once a group called **the alchemists** popped up. These folks didn't exactly have a hunger for pure knowledge, though – their aim was to try and turn basically anything they could into an element that still fetches a lot of money today: gold!

Unfortunately for them, however, they never did manage to crack the secret of alchemy, but their experiments did pique an interest into what things were actually made of. Once the 20th century finally rolled around, chemists had gone about the gargantuan task of answering this question and nearly all the elements had been discovered.

We've known about some of these for ages: metals like copper, gold, silver; and carbon too – which even in its elemental form is incredibly versatile, taking the form of anything from coal and diamonds, to the black stuff you'll find when you've burnt your toast. Then there are all the ones that don't really roll off the tongue, like darmstadtium – extremely radioactive stuff that we think is a kind of metal as it has things in common with other metals like platinum and nickel, but we're not entirely sure yet. Clearly, there's still more to learn.

Even when something is made out of just one type of atom, though, different configurations of that atom can still give an element different properties (which is why carbon takes so many forms). **Also, some atoms do just float about enjoying the single life, like our unreactive friend helium, while other atoms are more social.**

EDITOR'S NOTE) WHY IS EVERYTHING A SOCIAL ANALOGY CHARLIE?

However, many elements prefer going about in pairs, or even bigger groups of atoms of the same type – we call these packs of atoms 'molecules'. The oxygen that we breath mostly exists as molecules made up of two oxygen atoms, for example.

THE PERIODIC TABLE

All the elements we know about have been handily organized in a big chart called the Periodic Table of Elements – which might just seem like giant block of squares (because, well, it is) but it organizes the elements in some really clever ways. Given its complexity, you'd think it would have taken a long time to organize them all – however Dmitri Mendeleev apparently came up with the basis for the table in a dream! Kind of a boring thing to dream about?

So, how does the table work? Well, we know that the atoms that make up a single element – carbon, for example – are all the same. But what makes them the same? Well, every carbon atom has something very important in common: they each have the same number of protons (six). A proton is an even smaller type of particle that exists within the atom (more on them later) and the number of protons an atom has determines what element it is. This, in turn, denotes an atom's 'atomic number', which is basically the order that the elements appear in the table!

Hydrogen, the most abundant element in the whole universe, is also the simplest. It's got just the one proton, so its atomic number is 1, putting it first in line on the table. In contrast, gold has 79 protons, so its atomic number is 79. Nice and simple! However, the chart also groups elements with similar characteristics together – so, travel down the column under helium, and you'll find all the 'noble gases' which are similar to it. By listing all of the elements this way and spotting any gaps that have appeared, scientists have actually been able to predict what any undiscovered elements would be like, even before they had ever managed to make them in a lab.

CREATING THE ELEMENTS

While organizing the elements into a lovely pattern might be quite satisfying (and useful!), what it doesn't show us is how these elements came to exist in the first place.

Remember when I threw that whole "we are all made of stardust!" bombshell on you back in Chapter One? Well... if I hadn't gotten myself all excited to tell you about that right away, this is probably where that information would have ended up. Still! Let's recap...

After the Big Bang, the simplest and lightest of the elements were formed: the gases hydrogen and helium, along with traces of a few others. (This actually means that when you fill up a party balloon with helium, most of those atoms are almost as old as the universe itself!) Then, as the universe developed and the dust from the Big Bang gathered into stars and galaxies, more elements were formed in the heart of those stars where temperatures were just so hot that the nuclei (or centres) of the lighter, simpler elements fused to form new, heavier elements.

As stars get older, they're able to produce new, more complicated elements. When a star explodes it sends all of its material out into the universe; even then it's still making new elements, as the heat from that supernova can produce atoms like lead, platinum and gold.

The heaviest element you'll find in nature (in a significant amount, at least) is uranium, which has 92 protons in its nucleus. So why did nature stop there? Well, there's a natural limit to how big an atom can be. Eventually, you get to a point where there are so many protons jostling about in the nucleus that it sort of just starts overflowing. This is where radioactivity comes in. Over a long time, a heavy atom like uranium starts to undergo radioactive decay – releasing smaller particles and energy – until it becomes something else entirely. In uranium's case, that's lead.

If the heaviest element found in nature has 92 protons, how have we ended up with 118 of them? Well, it seems that when the stars stopped making new elements, people took the task into their own hands. In recent years, people have been playing around with atoms in machines called particle accelerators. So how do you use one of these to make a new element? Well, you simply take two atoms, fire them at extreme speeds and **SMASH** them together!

It's through this (seemingly haphazard, but actually very scientific, I'll have you know) technique we've been able to come up with new elements not found in nature. Well, you would have found elements like these when the Earth was young, but they were simply too unstable to stick around. These new, manmade elements are really big, and so they're really radioactive too, and quickly decay into smaller, more stable elements. For example, if you fire some neon atoms at the already radioactive metal curium, you end up with atomic number 106 – 'seaborgium' (I just checked and it was actually named after the scientist Glenn T. Seaborg but it's way more fun to imagine that it was actually named after a cyborg that lives under the sea) of which only a few atoms have ever been made.

INSIDE THE ATOM

For a long time it was thought that atoms were the smallest things that ever existed. The clue's in the name – 'atom' is Greek for indivisible, something that can't be split up any smaller... so you can imagine everyone's surprise when it was discovered that atoms actually are divisible! (Although you probably wouldn't want to divide them. **See: the atomic bomb.**)

Introducing... the subatomic particles! The world inside the atom. 'Atomic' from the Greek for 'indivisible', and 'sub' from the Greek for 'we totally got this one wrong guys LOL.' (I'm guessing.)

EDITOR'S NOTE THE GREEKS CERTAINLY DIDN'T USE THE TERMINOLOGY 'LAUGH OUT LOUD' LET ALONE THE ACRONYM 'LOL'.

At the heart of an atom you'll find its nucleus, where the protons and other tiny particles called neutrons jostle about, and it's these protons and neutrons that give an atom most of its mass. Then there are electrons, which are even smaller – so small in fact that, even with today's technology, we still can't measure them. These electrons orbit around the nucleus, sort of like how the planets orbit around the Sun.

A diagram of an atom usually looks something like this:

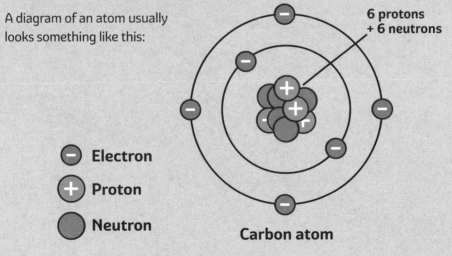

6 protons + 6 neutrons

Electron
Proton
Neutron

Carbon atom

But the big surprise here is the diagram on the previous page is in no way drawn to scale. If we were somehow able to make an atom the size of an entire football stadium, then the nucleus would still only be the size of a pea. So what lies between the nucleus and the electrons that are whizzing about the edge of the stadium?

Well, here's a scary thought: **NOTHING DOES.**

But actually, though, a hydrogen atom is, staggeringly, 99.9999999999996% empty space, and this goes for absolutely all of the matter in the universe too. AND YES, THAT INCLUDES YOU. **You are, technically speaking 99% nothing. (I'm not sure if I really planned it this way, but it seems like my main purpose of this book has been to try and induce as many existential crises as possible. My bad.)**

The electrons, circling around the nucleus, are attracted to the protons within it by a force called electromagnetism. Just like you're used to with a magnet, opposites attract! So, the negatively charged electrons are attracted to the positively charged protons. Neutrons on the other hand are (and the clue's in the name here) neutral, with no charge. In fact, because they don't make much difference, atoms of a single element can have different numbers of neutrons, and they'll still belong to that same element – these different varieties of the same element are called 'isotopes'.

So, how deep does this rabbit hole go? Are these electrons, protons and neutrons made of anything too? Can we just go on and on, cracking particles open forever? Well, so far, it seems electrons have no structure that we know of – so they're a dead end for now. Protons and neutrons, however, are made up of still more subatomic particles called quarks... but we'll get to those a bit later.

HOW EINSTEIN PROVED THAT ATOMS EXIST

While today we simply take it as scientific fact that all matter is made up of atoms, it definitely wasn't always that obvious. I mean, even with all of the knowledge we have today, it still can seem a little counter-intuitive. We're made up of particles that are far too small to see? And which are predominantly made of 99% NOTHING? How on earth do you prove that?

DON'T WORRY GUYS, I GOT THIS

Back in 1905, it was a young Albert Einstein who proved that atoms do, in fact, exist. Back then, there was still a huge question mark hanging over the whole concept. Many scientists thought that matter was probably made up of these little things called atoms, but they'd still never seen one of them.

Fortunately Einstein swooped in to set the record straight. **(In my head, he swooped in on a hang-glider, for whatever reason.)** How did he do it? Remember how with dark matter, even though we can't see it, we still know that it's there because of its effects? The effects that atoms create was what Einstein went about measuring...

JUST THINK, FOR A MOMENT, ABOUT WHAT HAPPENS WHEN BITS OF DUST ARE FLOATING ABOUT IN THE AIR. WHAT DO THEY LOOK LIKE? WELL, THEY'RE NOT EXACTLY STATIC — INSTEAD OF STAYING STILL, THE PARTICLES SEEM TO MOVE ABOUT IN A RANDOM, JITTERY WAY. WHY IS THAT? THE IDEA WAS THAT THESE TYPES OF MOVEMENTS AREN'T REALLY RANDOM AT ALL, BUT ARE ACTUALLY THE RESULT OF MUCH TINIER PARTICLES BETWEEN EVERYTHING, WHICH ARE BUMPING INTO EVERYTHING ELSE, AND THUS MAKE THE BITS OF DUST APPEAR TO SHAKE!

IT WAS EINSTEIN WHO SHOWED THAT IT WAS THE MOVEMENT OF TINY ATOMS BUMPING INTO EACH OTHER THAT WAS MAKING THE BIGGER DUST PARTICLES MOVE... AND EVEN BETTER, HE ALSO CAME UP WITH A WAY TO PREDICT MATHEMATICALLY EXACTLY HOW THE PARTICLES WOULD 'JITTER', AND HOW BIG THESE ATOMS ACTUALLY WERE. ULTIMATELY, THIS WAS WHAT THE SCIENTIFIC COMMUNITY NEEDED TO PROVE THAT THESE ATOM THINGS WERE DEFINITELY THERE.

THE GOLD FOIL EXPERIMENT

As it turns out, 1905 was a very big year for atoms! That same year, a scientist called Ernest Rutherford fired a beam of alpha particles (tiny, positively charged particles, each made of two protons and two neutrons) at a piece of very, thin gold foil. Why? Well for fun, obviously!

OK, he actually did have a scientific reason. The thought at the time was that, close up, atoms would look a bit like 'plum puddings' – basically a positively charged sphere filled with negatively charged electrons. Think of the positive charge like the pudding's crust, and the negative electrons like the fruit inside. (Clearly another case of scientists naming something when they were hungry.)

The idea was that, if this was the case, the alpha particles should pass in a straight line through the gold, as there was actually nothing there to stop them. However, to everyone's shock, not only did some particles go through and emerge at odd angles, but some of the particles bounced right back at the scientists! The particles were being repelled by something that the team didn't expect to find – the gold atoms' small but-not-to-be-ignored centres: their nuclei. Thus, the true configuration of the atom was discovered.

CHAPTER 9

THE

PARTICLE

NOW IT'S TIME TO GO BEYOND THE ATOM, TO THE SMALLEST OF THE SMALL. IN FACT, TO GIVE YOU AN IDEA OF SCALE, THE REST OF THIS CHAPTER WILL ONLY BE POSSIBLE TO READ WITH A MICROSCOPE!

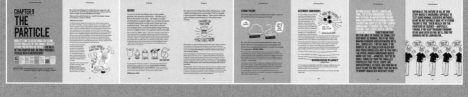

EDITOR'S NOTE ⟩ WE'RE ALSO TAKING THE LIBERTY OF PRINTING THIS
CHAPTER IN NORMAL SIZE TOO, FOR OBVIOUS REASONS.

OK, so this is where things get so small that they start to get a bit... odd. On this level, not only do the super-small components constitute what our universe is made of, but they also effect how it behaves.

Welcome to the world of the quantum – where for this chapter only, you are a mechanic... a quantum mechanic.

Why? Because I say so.

The thing about the subatomic universe is that it doesn't seem to play by the rules of the familiar, everyday world that we live in. For instance, usually science is very keen on pinning down exactly what's happening and where... but here, something called the 'uncertainty principle' means that you can't actually work out where a particle is and how fast it's moving at the same time. Not only that, but the idea that one particle exists in two different places at once is completely valid! So instead of using normal methods, you have to pin your hopes on probabilities, and embrace a bit of 'fuzziness' about the whole thing. Oh, and another thing: when you start looking at stuff at this minuscule level, it doesn't really work to describe things as simply being particles (like electrons) or waves (like light) because, up close, things look and behave like both waves and particles at the same time!

It's almost as if, at this point, scientists had absolutely no idea what was going on anymore... and so they just started making up stuff like the 'uncertainty principle' so that it would seem like being uncertain of everything was what they were planning all along.

Regardless of how counter-intuitive all of this stuff might seem, though, I promise it's **DEFINITELY STILL SCIENCE**. So, with all that in mind, here's what we know so far about...

QUARKS

We can't break the electron down into smaller particles, but neutrons and protons – the bits you'll find at the centre of an atom – are made of smaller particles known as quarks. Right now, we think of quarks as being fundamental, so they aren't made up of any smaller particles. Then again, given that scientists once thought atoms were the smallest things knocking around, it might be best not to get too used to this idea.

Quarks come in six different 'flavours'. (Yes, that's how they actually refer to them, although there is no scientific evidence that shows how they might taste). This is just a way of dividing up the six different kinds of quark, which are called – wait for it – up, down, charm, strange, top and bottom. The last two actually used to be called 'truth' and 'beauty', but apparently scientists decided that was too cheesy. Quarks also come in three different colours – again, these aren't colours we can see, just a way of dividing them up – which are blue, red and green.

UP   DOWN   CHARM   STRANGE   TOP & BOTTOM

A group of quarks is known as a hadron – and you'll always find quarks in groups like these, never out on their own. **(I mean, they're pretty little, so fair enough that they'd want to stick together.)** Protons and neutrons, in fact, are hadrons: a proton is made up of two 'up' quarks and one 'down' quark, while a neutron is made of one 'up' and two 'down' quarks.

With me so far?

OK, just one more thing. Remember the electron, which couldn't be broken down any further? Well it's actually a type of particle called a lepton. There are six different types of lepton, but the electron is the best known of them.

CONGRATS ON YOUR *FAME*

MR. ELECTRON

While quarks are always knocking about with other quarks, leptons are loners. I know this is a lot of bits and pieces to get your head around if you're not already familiar with all of these fundamental particles, but what you can take away is this: **pretty much everything around us is, when you break it down, made of leptons or quarks.**

And one tiny little final point to mention: for every type of matter particle, there's also an antimatter version – a particle that behaves in the same way, except that it has an opposite electrical charge. So the antimatter version of an electron, which has a negative charge, is the positron, which has a positive charge. Oh, as well as quarks, there are anti-quarks – **you can think of this stuff like matter's secret, weird twins that they don't like anyone to know about.**

Weirdly, though, while you might assume that the universe would have equal amounts of both matter and antimatter swirling about, there does seem to be much more matter than antimatter (which is handy, or else the two things would have cancelled each other out, and we wouldn't be here today). We don't know exactly why this is, but it does all feed into...

STRING THEORY

Here's an idea that has revolutionized modern science: the whole universe is made up of string!

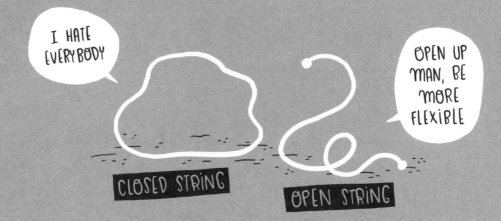

Ok, well, it's a bit more complicated than that... hey, don't mind me. I'm just going to try and explain one of the most complicated scientific concepts in a few paragraphs. **NO BIG DEAL.**

So, now you know a bit about the elements, the atoms, and the subatomic particles that make up atoms. However, what string theory says is that, when you think about those tiniest particles you shouldn't be imagining them like the perfect spheres you'll see in school textbooks. Instead, string theory proposes the idea that those fundamental particles are like tiny little vibrating filaments of energy – strings, basically.

Instead of having different strings representing particles (which would be too simple for quantum mechanics) what happens is that, depending on how they're vibrating, different strings make different particles! It's actually quite a neat, elegant way of looking at the world... but there's one catch to it all.

To make the theory work, the universe would need to have many more dimensions than the ones that we're used to – otherwise, the calculations behind the whole system wouldn't work.

ALTERNATE DIMENSIONS

As human beings, living day-to-day on Earth, we're aware of moving about in the three dimensions of space: going left and right; back and forth; up and down – 3D, essentially. There is also a fourth dimension, which is equally crucial to how our sense of the world works: time. Our universe is actually 4D!

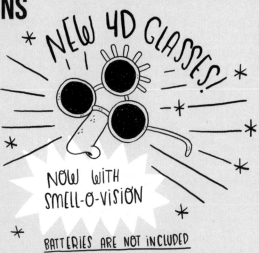

Or at least that's the way that we perceive it, but it's possible that we might have undershot the true number of dimensions by quite a bit. Why do we think this? Well, if we had a few more hidden dimensions to work with, then it'd go a good way to explaining some of the slightly more odd things that the universe does.

For example, there are various forces that affect the way particles behave. Electromagnetism is one, which is the force that makes a magnet stick to a metal door. The magnet is also affected by the force of gravity – the attraction seen between not just objects and the Earth, but between any two objects (and the more massive an object, the stronger its pull).

The funny thing is, gravity doesn't seem to be as powerful as it should be – in this example, if you hold the magnet close to the metal door, it can still 'jump' across the space and stick itself to the door... even given the gravitational force of an **ENORMOUS PLANET** **pulling it down.**

Why is gravity so apparently weak? Well, the answer might be that we're simply not feeling the whole force of gravity, as it's also acting in other dimensions that we don't experience.

And why don't we notice these dimensions? Well, they could just be really, really small.

BEYOND HEIGHT, WIDTH, LENGTH AND TIME, THE OTHER DIMENSIONS (THERE ARE 10 IN SUPERSTRING THEORY, AND A WHOPPING 26 IN BOSONIC STRING THEORY) ARE SORT OF LIKE TINY 'LOOPED' DIMENSIONS. IT WORKS LIKE THIS: LET'S IMAGINE THAT LENGTH WAS A 1M LOOPED DIMENSION. THAT'D MEAN THAT WE'D BE ABLE TO TRAVEL UP, DOWN, LEFT AND RIGHT AS NORMAL, BUT IF WE TRIED MOVING FORWARDS AND BACKWARDS WE'D SEE... OURSELVES, 1M AWAY. IF WE WANTED TO, WE COULD EVEN REACH OUT AND TOUCH OURSELVES! (NOT IN THAT WAY.) THE OTHER, HIDDEN DIMENSIONS WOULD WORK LIKE THIS — HOWEVER, THEY'RE SO SMALL (SMALLER THAN THE SMALLEST PARTICLES) THAT THESE 'LOOPS' ARE IMPERCEPTIBLE. AS SUCH, THE FOUR WE'RE USED TO ARE THE ONLY ONES THAT WE NEED TO WORRY OURSELVES WITH DAY TO DAY.

NATURALLY, THE NATURE OF ALL OF THIS STUFF MAKES IT INCREDIBLY DIFFICULT TO TEST USING NORMAL SCIENTIFIC METHODS... SO WE'RE NOT ENTIRELY SURE YET IF STRING THEORY IS TRUE. THESE REALLY ARE THE FRONTIERS OF SCIENCE — HOPEFULLY, THOUGH, IF WE KEEP TRUDGING FORWARD AS WE HAVE BEEN SO FAR, WE'LL FIND THE ANSWERS WE'RE LOOKING FOR...

CHAPTER 10

THE END

OF TIME

(AND OF THE BOOK)

WELL, I GUESS IT'S TIME FOR ME TO GO.

I'M RUNNING OUT OF TIME, YOU SEE! WHICH INHERENTLY MEANS I'M RUNNING OUT OF SPACE, TOO. YOU CAN'T SEPARATE THE TWO, OF COURSE – IT WAS EINSTEIN WHO REALIZED THAT THEY ARE INEXTRICABLY LINKED, IN WHAT WE CALL THE SPACE-TIME CONTINUUM.

THERE ARE A FEW THINGS I COULD TRY TO SLOW THINGS DOWN: GRAVITY ACTUALLY DISTORTS TIME, SO THE STRONGER THE GRAVITATIONAL EFFECT IS, THE SLOWER TIME TRAVELS. THE BEST WAY FOR ME TO SLOW TIME DOWN WOULD PROBABLY BE TO JUMP INTO A BLACK HOLE, WHERE THE GRAVITATIONAL FORCE IS SO STRONG THAT TIME WOULD BASICALLY SLOW DOWN TO A STOP... BUT I COULDN'T EXACTLY DO MUCH WRITING INSIDE OF ONE OF THOSE THINGS (I'd be very dead you see).

Moving faster through space actually slows down time too – we've proved this by sending clocks off on spaceships, which come back ticking fractions of a second behind their Earth-bound counterparts. Again though, the only way for this to have any meaningful effect is to travel at nearly the speed of light and no vehicle invented so far is quite that quick just yet.

Given time's apparent malleability, it's no wonder that in this weird, wonderful universe, **time travel no longer seems an impossibility.** There's always a chance that we might discover wormholes – bridges through space and time – they're not impossible! But, while physics might allow their existence, they'd take insane amounts of energy to keep open, so it's very unlikely we'd be able to travel through one.

Thus, while I travel along the earthly timeline like everyone else, this **Fun Science** journey must come to an end. But is the end of time really the end? Or is it just a new beginning?

I guess it's time to find out,

THE END.

EDITOR'S NOTE I CAN'T TELL IF THAT WAS A PHILOSOPHICAL COMMENT ON THE CYCLICAL NATURE OF THE UNIVERSE, OR IF YOU'RE JUST HINTING AT THE FACT THAT YOU WANT ANOTHER 'FUN SCIENCE' BOOK DEAL? IN EITHER CASE: WELL PLAYED, CHARLIE. WELL PLAYED INDEED.

INDEX

ACKNOWLEDGEMENTS

A project of this scale doesn't happen in a vacuum. That's because there's no air in a vacuum, so I wouldn't be able to breathe in one, and without respiration my body wouldn't be able to convert the food I eat into energy, and that'd ultimately result in my death. So, vacuums aren't exactly conducive environments for good writing.

However, this book also wasn't created in isolation. Yes, most of the work I did on the book happened alone at my desk, with head-phones in, listening to 'Lifeformed's' *Fastfall* album over and over again – sometimes late into the night and often on weekends to get everything finished on time. But, while it might be my silly mug and name sitting on the cover, creating *Fun Science* really was a huge collaboration.

First and foremost I'd like to thank my editor, Romilly, not only for all of the contributions you made to the text (a lot of which involved removing my many, many digressions) but also for the amazing patience and caring you showed me while I consistently missed my deadlines and huddled in the corner hoping you'd forget about me. (Pro tip: That technique never works.) Also, thanks for letting me write you in as a character in your own right – don't be fooled folks, Romilly is actually quite lovely.

Acknowledgements

Thanks to all the folks at Quadrille for believing in this project from the start, and for publishing a very different kind of YouTuber book, including: Sarah, Helen, Caroline, Inez and Margaux. To Dave and Fran for your absolutely incredible designs and illustrations respectively – you have both really given this book a style and life that I think is truly unique as far as science books go. To Jenny, for all your hard work collecting enough facts and research for me to fill about ten books; to my last minute writing consultant Emma for playing an invaluable part in getting the book together and to Emily, Simon, Tilly, and the other fine folks at Storm Management for sticking by my side through this whole shebang.

Thanks to Mum, Dad, Will and Bridie for always supporting me and this weird internet thing that I do; to Emily for all your love, for keeping me sane throughout this process, and for being there to pat me on the back every time I handed something in (and for reminding me to pat myself on the back too).

Most importantly though, I'd like to thank... you! And everyone else who's supported my 'Fun Science' endeavours over the years. There's absolutely no way I would have put fingers to keyboard if it hadn't been for all of the comments, tweets and real-life gratitude you've shared with me about how my videos (and now hopefully this book too) have made you see science in a new way. I cannot overstate how important you all have been throughout this process – you lot are absolutely the reason that this book was able to be made in the first place. So again, thank you, thank you and thank you!

All my love,

Charlie ☺

THE HUBBLE EXTREME DEEP FIELD

If you point a powerful telescope at a blank area of sky and leave it there for a while, this image is what you get – just a small taster of the millions of galaxies that lie out there in the universe...

EDITOR'S NOTE   *WHAT FEELING DO YOU WANT PEOPLE TO LEAVE THIS BOOK WITH CHARLIE? INSIGNIFICANCE?*

Well, that's the universe for you...